ACE YOUR MIDTERMS & FINALS

FUNDAMENTALS OF MATHEMATICS

Other books in the Ace Your Midterms and Finals Series include:

Ace Your Midterms and Finals: Introduction to Psychology

Ace Your Midterms and Finals: U.S. History

Ace Your Midterms and Finals: Principles of Economics

Ace Your Midterms and Finals: Introduction to Physics

Ace Your Midterms and Finals: Introduction to Biology

ACE YOUR MIDTERMS & FINALS

FUNDAMENTALS OF MATHEMATICS

ALAN AXELROD, PH.D.

McGraw-Hill
New York   San Francisco   Washington, D.C.   Auckland   Bogotá
Caracas   Lisbon   London   Madrid   Mexico City   Milan
Montreal   New Delhi   San Juan   Singapore
Sydney   Tokyo   Toronto

Library of Congress Catalog Card Number: 99-070654

McGraw-Hill

A Division of The McGraw·Hill Companies

Copyright © 1999 by The McGraw-Hill Companies, Inc. All rights reserved. Printed in the United States of America. Except as permitted under the United States Copyright Act of 1976, no part of this publication may be reproduced or distributed in any form or by any means, or stored in a database or retrieval system, without the prior written permission of the publisher.

1 2 3 4 5 6 7 8 9 0 DOC/DOC 9 0 3 2 1 0 9

ISBN 0-07-007008-3

The sponsoring editor for this book was Barbara Gilson, the editing supervisor was Maureen B. Walker, the designer was Stateless Design for The Ian Samuel Group, Inc., and the production supervisor was Tina Cameron.

Printed and bound by R. R. Donnelley & Sons Company.

McGraw-Hill books are available at special quantity discounts to use as premiums and sales promotions, or for use in corporate training sessions. For more information, please write to the Director of Special Sales, McGraw-Hill, 11 West 19th Street, New York, NY 10011. Or contact your local bookstore.

 This book is printed on recycled, acid-free paper containing a minimum of 50% recycled, de-inked fiber.

PART FOUR
FOR YOUR REFERENCE

HOW TO USE THIS BOOK

YOU KNOW THE DRILL. FIRST DAY IN A SURVEY, INTRODUCTORY, OR CORE COURSE: THE professor talks about grading and, saying something about the value of the course in a program of "liberal education," declares that what he or she wants from students is original thought and creativity and that, above all, he or she does not "teach for" the midterm and final.

Nevertheless, the course certainly *includes* one or two midterms and a final, which account for a very large part of the course grade. Maybe the professor can disclaim with a straight face *teaching for* these exams, but few students would deny *learning for* them.

True, you know that the purpose of an introductory course is to gain a useful familiarity with a certain field and *not* just to prepare for and do well on two or three exams. Yet the exams *are* a big part of the course, and, whatever you learn or fail to learn in the course, your performance as a whole is judged in large measure by your performance on these exams.

So the cold truth is this: more than anything else, curriculum core courses *are* focused on the midterm and final exams.

Now, traditional study guides are outlines that attempt a bird's-eye view of a given course. But *Ace Your Midterms and Finals: Fundamentals of Mathematics* breaks with this tradition by viewing course content through the magnifying lens of ultimate accountability: the course exams. The heart and soul of this book consists of midterms and finals prepared by *real* instructors, teaching assistants, and professors for *real* students in *real* schools.

Where did we get these exams? Straight from the professors and instructors themselves.

◆ All exams are real and have been used in real courses.

◆ All exams include critical "how-to" tips and advice from their creators and graders.

◆ All exams include actual answers.

Let's talk about those answers for a minute. In most cases, the answers are actual student responses to the exam. In all cases, the answers included are A-level responses and you'll find full commentary—provided by the instructors—that points out why the response is successful.

This book also contains more than the exams themselves.

◆ In Part One, "Preparing Yourself," you'll find how-to guidance on what math professors look for, thinking like a mathematician, studying more effectively, and gaining the performance edge when you take an exam.

◆ Part Two, "Study Guide," presents a quick and easy overview of the content of the three most common and popular core math courses: algebra, trigonometry, and calculus. This part of the book clues you in on what to expect in these courses.

◆ Part Three, "Midterms and Finals," contains the exams themselves, grouped by subject (algebra, trigonometry, and calculus) and by college or university.

◆ In Part Four, "For Your Reference," you'll find a handy guide to common math symbols and a brief list of recommended reading.

What This Book Is Not

Ace Your Midterms and Finals: Fundamentals of Mathematics offers a lot of help to see you through to success in these important courses. But (as you'll discover when you read Part One) the book is *not* a math skills review and certainly *cannot* take the place of

◆ Doing the assigned reading

◆ Working assigned homework problems

◆ Working extra practice problems

◆ Keeping up with your work and study

◆ Attending class

◆ Taking good lecture notes

◆ Thinking about and discussing the topics and issues raised in class and in your books

Ace Your Midterms and Finals: Fundamentals of Mathematics is not a substitute for the course itself!

What This Book Is

It's both cynical and silly to invest your time, brainpower, and money in a college course just so you can ace a couple of exams. If you get A's on the midterm and final, but come away from the course having learned nothing, you've failed.

We don't want you to be cynical or silly. The purpose of introductory, survey, or core courses is to give you a panoramic view of the knowledge landscape of a particular field. The primary goal of the college experience is to provide more than tunnel intelligence. It is to enable you to approach whatever field or profession or work you decide to specialize in from the richest, broadest perspective possible. College is education, not just vocational training.

We don't want you to study for the exam: the idea is to study for the rest of your life. You are buying knowledge with your time, your brains, and your money. It's an expensive and valuable commodity. Don't leave it behind in the classroom at the end of the semester. Take it with you.

Even the starriest-eyed idealist can't deny, however, that midterms and finals are a big part of intro courses and that even if your ambitions lie well beyond these exams (which they should!), performing well on the tests is necessary to realize those loftier ambitions.

Don't, however, think of midterms and finals as hurdles—obstacles—you must clear in order to realize your ambitions and attain your goals. The exams are there. They're real. They're facts of college life. You might as well make the most of them.

Use the exams to help you focus your study more effectively. Most people make the mistake of confusing *goals* with *objectives.* Goals are the big targets, the ultimate prizes in life. Objectives are the smaller, intermediate steps that have to be taken to reach goals.

Success on midterms and finals is an objective. It is an important, sometimes intimidating, but really quite doable step toward achieving your goals. Studying for—working toward—the midterm or final is not a bad thing, as long as you keep in mind the difference between objectives and goals. In fact, fixing your eye on the upcoming exam will help you to study more effectively. It gives you a more urgent purpose. It also gives you something specific to set your sights on.

And this book will help you study for your exams more effectively. By letting you see how knowledge may be applied—immediately and directly—to exams, it will help you acquire that knowledge more quickly, thoroughly, and certainly. Studying these exams will help you to focus your study in order to achieve success on the exams—that is, to help you attain the objectives that build toward your goals.

CONTRIBUTORS

Ryan P. Albert, *Graduate Teaching Assistant, Mathematics, Ohio State University*

Fedor A. Andrianov, *Lecturer in Mathematics, University of Chicago*

Jonathan Andrew Bodrero, *Graduate Student Instructor, Department of Mathematics, Brigham Young University*

Stephanie L. Fitch, *Lecturer in Mathematics, University of Missouri-Rolla*

Mark A. McCombs, *Director of Teacher Training, Department of Mathematics, University of North Carolina*

Nephi Noble, *Graduate Student Instructor, Department of Mathematics, Brigham Young University*

Jodi Petersime, *Associate Instructor of Mathematics, Indiana University*

Emma Previato, *Professor of Mathematics, Boston University*

Rasul Shafikov, *Associate Instructor of Mathematics, Indiana University*

Jennifer C. Stevens, *Instructor in Mathematics, University of Tennessee*

ABOUT THE AUTHOR

Alan Axelrod, Ph.D., is the author of numerous books, including *Booklist* Editor's Choice *Art of the Golden West, The Penguin Dictionary of American Folklore,* and *The Macmillan Dictionary of Military Biography.* He lives in Atlanta, Georgia.

PART ONE

PREPARING YOURSELF

CHAPTER 1

FUNDAMENTALS OF MATH: ONE SIZE DOES NOT FIT ALL

UNIVERSITY AND COLLEGE MATH DEPARTMENTS OFFER MANY DIFFERENT MATH courses taught by a variety of teaching assistants, instructors, and professors, who may bring to their subject a broad spectrum of approaches, attitudes, and teaching philosophies. So let's begin by defining the scope of this book and explaining how it can help you excel in the core or fundamental math course you are taking.

Is It Fundamental?

If math departments, courses, and teachers come in more flavors than a certain well-known chain has of ice cream, math students—especially nonmajors—are even more varied in the backgrounds they bring to class, their math skills, their aptitude for math, their attitudes toward math, and their reasons for taking a particular course.

Background knowledge may range from woefully inadequate to quite strong. The range of skill levels is commensurate with this range of background. As for aptitude for the subject, that's at least as varied as what the roll of genetic dice produces. Attitudes? They range from fear and loathing to genuine love for the subject—and everything in between. For many students, however, math is a tool—no more, no less—a means to an end, a necessary adjunct to a major or to career plans. This also touches on the variety of reasons students have for taking math:

◆ It may be a general curriculum core requirement.

◆ It may be required as an adjunct to a major (economics, say, or one of the sciences).

◆ It may be that the student believes (quite correctly) that a fundamental knowledge of math is an important part of his or her education.

◆ It may even be that the student, though not a math major, *likes* math—is *interested* in it.

All of this variety means that the odds are that one course the math department in your university or college does *not* offer is something specifically titled "Introduction to Mathematics," or "Survey of Mathematics," or even "Fundamentals of Mathematics."

Why not?

The short answer is that mathematics is too broad a field to make such a course useful or practical on the college level, especially in the span of a single semester.

The even shorter answer is that what constitutes an adequate, appropriate, and useful introduction, survey, or fundamental course for one student will be too broad, too narrow, too basic, or too advanced for another.

> Some math departments offer discrete mathematics as an introductory course. This branch of mathematics deals with abstraction, notation, critical thinking, and logic structures, especially as these relate to computer and information sciences. With the rise of the computer, discrete mathematics has assumed increasing importance, and, on an introductory level, may be taken as preparation for computer science. At this point, however, most students who take discrete mathematics are relatively advanced. It has yet to enter the mainstream on an introductory level.

A Math Menu: The Starters

What, then, is a *fundamental* math course?

In the overwhelming majority of American colleges and universities, three types—or levels—of math courses are offered as basic:

◆ Algebra

◆ Trigonometry

◆ Calculus

This is obviously a short menu. Of course, other areas of math—geometry, for example—are basic, too. And all math departments offer introductory geometry courses, as well as introductory courses in other areas.

It is also true that there are topics in math that most of us would consider more basic than these. What about pre-algebra—basically, the more advanced kinds of arithmetic? It is true that most math departments offer review or remedial pre-algebra courses for students whose algebra skills are rusty or nil. These courses, however, are usually not counted toward fulfillment of core curriculum requirements. Many professors of mathematics also recognize that our culture tends to be anti-math at worst, or to seriously underprepare students in mathematics. Accordingly, some math departments offer core courses in "numeracy"—which is to mathematics what literacy is to language. Numeracy courses are the closest most math departments come to courses that might adequately be labeled "Fundamentals of Mathematics." These courses generally stress concepts over computation and seek to create a math awareness and "math culture" among students.

Philosophically, all of these approaches, then, may be considered approaches to the fundamentals of mathematics.

But *Ace Your Midterms and Finals: Fundamentals of Mathematics* is intended as a practical aid for the math student, not a philosophical discussion of the subject, and the approach we take here is to offer study help and test-taking guidance in those three most popular fundamental areas: algebra, trigonometry, and calculus.

Fundamental? To the student who struggles with basic college algebra, trigonometry and calculus hardly seem "fundamental" at all.

In the first place, *fundamental* is not synonymous with *simple.* And, yes, both trigonometry and calculus are conceptually and technically demanding. But they *are* fundamental to mathematics in and of itself as well as to mathematics as applied in engineering, the sciences, and (in the case of calculus) even business and economics. Depending on the background and goals of the student, algebra may be the appropriate introductory college math course. It may even be the only college math course a particular student takes. But, for many students, especially those preparing for majors and/or careers in the sciences, in many business-related fields, and in engineering-related fields, survey-level courses in trigonometry or calculus are the appropriate introductory math courses.

This book attempts to meet the needs of students who begin college math at any of the three most frequently encountered levels. To be sure, of these three levels, algebra is the most popular, so more of *Ace Your Midterms and Finals: Fundamentals of Mathematics* (Chapters 12–17) is devoted to these courses than to trigonometry (Chapters 18 and 19) and calculus (Chapters 20–22). However, even if you are in the majority and are concerned exclusively with algebra as your fundamental math course, you will benefit from reading the introductory material on trig and calculus and looking at the sample exams included for those subjects. Doing so will help you gain a better perspective on algebra and give you some idea of where in math you might go from there.

CHAPTER 2

WHAT MATH PROFESSORS WANT FROM THEIR STUDENTS

T'S OBVIOUS THAT THE EXPECTATIONS OF INSTRUCTORS TEACHING AN ALGEBRA SURVEY differ from those of instructors teaching first courses in trig or calculus. It's also a given that different teachers take different approaches. Simply put, some demand more than others. This said, there is something quite basic that all college-level math instructors *don't* want from their students.

They don't want arithmetic.

The study of mathematics begins in elementary school (and often even continues in high school) with arithmetic, which Peter Hilton, Distinguished Professor of Mathematics Emeritus at the State University of New York, Albany, has described as a "horrible, wretched subject, far removed from real mathematics." Arithmetic, says Professor Hilton (in his foreword to Jan Gullberg's *Mathematics: From the Birth of Numbers*), is filled with "boring, unappetizing algorithms and pointless drill-calculations," the avoidance of which is "perfectly natural and healthy."

Perhaps Professor Hilton overstates the case

Primary course objectives are to:
1. Help students to overcome the "math fear" and help them learn and excel.
2. Teach the basic concepts of algebra.
3. Encourage students to continue learning about mathematics by taking more advanced courses.

In addition to overcoming the math fear (it's really not all that bad!), I want my students to
1. See that math (algebra) is used in the real world and that it is worth their time to learn it.
2. Learn basic concepts of algebra: root-finding, definitions, handling story problems, etc.
3. Develop the ability to think abstractly (although this is only limited abstraction).

—Jonathan Andrew Bodrero, Graduate Student Instructor, Math 110: Algebra, Brigham Young University

against arithmetic, but the essential point is crucial nevertheless. While some high school math courses are indeed quite advanced, for the majority of students the transition from high school to college math courses is a shift from a calculation orientation to a concept orientation.

Now, make no mistake. All introductory-level algebra, trig, and calculus courses involve extensive calculation and require memorization of special terminology, theorems, and formulas. However, most instructors put greater emphasis on acquiring a truly *fundamental* grasp of the concepts that underlie the calculation. This does not give you license to gloss over or avoid working problems or to content yourself with a "general understanding" of this or that concept. The fact is that mathematical concepts can be expressed in words only to a limited degree. Somewhat less limited is their expression as formulas. The fullest expression of a mathematical concept comes in actually working problems based on the concept. Just be sure to keep means and ends in proper perspective. The arithmetical approach stresses the mechanics of working through the problem and arriving at the solution. The mathematical approach, in contrast, puts the emphasis not on the mechanics of arriving at the solution, but on how the mechanics and the solution illustrate the mathematical principles that underlie them.

There are exceptions to this stress on concept and principle.

◆ While most college-level math instructors emphasize concept and principle, some do so less strongly than others.

◆ Some concepts, especially in calculus, involve profound matters of theory that are better tackled by advanced students in advanced courses. Nevertheless, methods of calculation in some of these cases may be readily and commonly taught. In such instances, the teacher may purposely stress memorizing technique rather than exploring the underlying concepts.

◆ Even instructors who savor teaching concepts recognize that poor skills or faulty technique get in the way of acquiring conceptual understanding. While keeping conceptual understanding in the foreground, these instructors typically also insist on students working many practice homework problems in order to acquire the mathematical (and arithmetical) fluency that will make it possible to grasp the underlying principles firmly.

> **Note that many intro-level instructors ease the burden on brute memory by furnishing formula sheets with the exams they give. You still have to recognize which equations to use in a given situation, but the details are furnished for you. Instructors who furnish such memory aids are not just being nice, but are demonstrating their belief that learning concepts—really understanding them—is more worthy of your time and attention than acquiring a set of formulas by rote, with or without much comprehension.**

> **Course objectives include: first, to teach a set of mathematical skills that can be applied in later math courses; second, to develop the students' ability to problem solve; third, to motivate the students as far as math goes and for learning in general; and, fourth, to build character.**
>
> **I hope that students will leave this course with confidence in their ability to solve problems, math-related or not. I want my students to learn the important facets of algebra, which are required for any math courses the students may take beyond this one. Finally, students should be able to defend their thinking and be able to express this defense.**
>
> —Nephi Noble, Graduate Teaching Assistant, Math 110: College Algebra, Brigham Young University

Math: Applied and Pure

It is also likely that your instructor has a dual vision of his or her subject, which you may or may not share. Mathematicians speak of "applied" and "pure" mathematics. Applied mathematics regards the discipline as a means to an end. Calculus is required to solve many problems in physics and even in business. Trigonometry is indispensable to the engineer. Pure mathematics, in contrast, sees the discipline as something worth doing in and of itself. You don't look at a painting by van Gogh or listen to a symphony by Mozart and then ask, "What good is that?" As civilized human beings, we recognize art and music—as well as literature and sport—as things that are worth doing, period. The pure approach to mathematics puts the discipline on a cultural level with art, music, literature, and sport. Many students, especially non-math majors, have trouble seeing the subject this way. Perhaps it's that all-too-prevalent habit we have of confusing arithmetic with mathematics. By definition, arithmetic must be useful *for some other pursuit*. Mathematics may be similarly useful, but it is also a science in its own right without the need for validation by proving its usefulness.

Accept the dual view of math—as a pure discipline also capable of application to other disciplines—and you will take a significant step toward getting in tune with your instructor's attitude and expectations.

A Stepwise Progress

You don't have to be a natural-born math whiz to excel in mathematical subjects, but you do have to commit yourself to achieving *understanding* of the concepts the math course presents. Rote memorization of terms, theories, and formulas is not sufficient. You need to make a kind of contract with yourself to study, work practice problems, and ask questions in order to embrace each concept presented. For, above all else, mathematics requires the self-discipline to master concepts step by step and to ensure that each step is thoroughly understood before moving on to the next.

This step-by-step approach cannot be emphasized strongly enough. Step back and think about, say, algebra, from a big-picture perspective. What leaps out at you is the degree to which one concept relates to another and, indeed, is built upon another. If you miss one concept, one step, the next concept will be difficult or even impossible to understand fully.

My course is designed to build study skills for future mathematics courses and to prepare students for brief calculus or finite mathematics. The course examinations, three midterms and a final, include problem solving as well as multiple-choice questions that test

1. **Ability to solve equations and inequalities individually and in systems**
2. **Basic understanding of exponential and logarithmic functions**
3. **Ability to graph conic sections, solutions for systems of equations, and logarithmic and exponential functions**
4. **Ability to manipulate rational expressions accurately and efficiently**

—Jodi Petersime, Associate Instructor, COAS J112: Introduction to College Mathematics, Indiana University

Math 125 is designed to introduce and explore the calculus of algebraic, exponential, and logarithmic functions. The objective of the course is to familiarize the student with the basic concepts and techniques of differential and integral calculus and their applications in problem solving.

An intuitive understanding of concepts is stressed over theory and rigorous proofs. The graphing calculator is used to help the students think about the geometric and numerical meaning of calculus and to approximate numerical solutions to realistic application problems.

Topics include: limits; continuity; derivatives; techniques of differentiation; marginal analysis; curve sketching and optimization; definite and indefinite integration; integration by substitution; the Fundamental Theorem of Calculus; area between curves; applications of the integral; and numerical integration

—Jennifer C. Stevens, Instructor, Math 125: Basic Calculus, University of Tennessee

There is no substitute for sound study habits. It is especially important to:

- **Do homework assignments every day.** A sure way to falter in mathematics courses is to allow yourself to fall behind in assignments.
- **Read each section and try the assigned problems prior to lecture.**
- **Get help when you don't understand a step, problem, or topic. If you wait until the exam, it's too late!**
- **Ask questions in class.** Chances are, others have the same question. Don't hesitate to ask.
- **Take full advantage of all the help that is offered, including office hours and free help sessions.**

You should make a friend and do your homework with someone. This allows you to check your work and methods; however, prepare for all exams and quizzes alone, on your own. Avoid using a calculator to prepare for an exam if the use of a calculator will not be allowed in the actual exam.

Do practice problems. Math is not a spectator course. You cannot learn it through osmosis.

—Jodi Petersime, Associate Instructor, COAS J112: Introduction to College Mathematics, Indiana University

A common mistake students make in a calculus course is putting stress on the formal memorization of isolated words (not ideas) and isolated techniques. Understanding of the general picture is often lacking, which results in the inability to solve problems that are even slightly different from those detailed in class. My advice is to put the stress on *understanding the material.* Calculus is not poetry!

—Fedor A. Andrianov, Lecturer Mathematics 350-152: Calculus, University of Chicago

Your instructor expects you to commit to attending class, to reading your book, and to doing the assignments. Beyond this, he or she expects that you will commit the time and concentration required to grasp each step in the mathematical process as it is presented. While it is important to keep up with the lectures and the progress of the class, it is also essential that you do so without cutting corners. You have to keep up, *but* you must not attempt to move from concept A to concept B without ensuring that you fully understand concept A.

Even in the simplest arithmetical problem, errors are cumulative. The situation with building mathematical understanding is analogous to this. Miss a step, and you will become increasingly bewildered until you are altogether lost.

The Work Ethic

Even for students who like math, algebra (or trig, or calculus) at the college level presents some formidable challenges. New concepts, new terms, and key equations must be learned. Now, as mentioned in the preceding chapter, many instructors, in an effort to put the emphasis on concepts rather than brute memory, do not demand that fundamental equations and formulas be committed to memory. On exams, they will include formula sheets, which list the basic equations you'll need to solve the exam problems. However, you still need to know which formulas to choose from this "kit," and that means connecting concepts with appropriate equations.

- Commit yourself to memorizing terms, concepts, and fundamental equations.

- Associate a concept with the appropriate mathematical expression and vice versa. Don't try to learn a list of concepts or equations in a vacuum.

- Similarly, it is best to learn associated terms, concepts, and equations together. For example, the time to learn about the concepts of domain and codomain is when you are learning about mapping.

- Identify and pay special attention to the key terms, theorems, concepts, and formulas. Your instructor will point out many of these. Obviously, too, if a term, concept, or equation is emphasized in lecture or in your textbook, you must regard it as key, and you should become thoroughly comfortable with it.

◆ The best way to learn new terms, theorems, concepts, and formulas is to *apply* them. Practice. Indeed, the only way to *learn* math is to *do* math.

◆ Focus on concepts, but continually develop your technique. Sound problem-solving habits will save you from making mistakes. Take the time to draw clear diagrams and to work calculations neatly. Make sure number columns align properly. Approach problem solving in an orderly manner.

Practice. Math courses don't just require brainpower. They are not just a set of concepts. Much of the work in math is calculation and manipulation of formulas. The more you practice applying the appropriate formulas to the appropriate problems, the easier day-to-day math becomes. You just have to work at it.

Study Habits

In addition to practice—which is especially important in all math courses—the study habits that work for other college courses work well for algebra, trig, and calculus, too. Because each lecture introduces new concepts, theorems, terms, and equations, it is very important to:

◆ Read textbook assignments *before* lectures.

◆ Keep up with your reading.

◆ Work homework problems *before* the lecture.

◆ Do lots of practice problems in addition to those assigned.

◆ Attend the lectures.

◆ Take good notes.

Visual Learning

Diagrams are often important in algebra and are indispensable in trigonometry and calculus. Those that you will encounter in your textbook and in lecture presentations are obviously more than decorative illustrations. They are important tools for understanding major concepts and are often critical to finding solutions to assigned problems. Get into the habit of making neat sketches that are accurate enough to reflect reality. A carelessly drawn and utterly improbable triangle will *not* help you solve a difficult trigonometry problem.

It's important to understand the theory given in class: basic definitions, statements of main theorems, etc. But understanding the theory, or "getting the idea," is not enough. You should be able to solve the problems, and solve them effectively. You cannot learn math by watching your classmate or an instructor. You have to do the work yourself. Practice: that's the most important ingredient in preparation for the test. Do a lot of sample problems, especially on the topics that you are least comfortable with. It's better to solve one problem fully understanding the solution than to solve ten problems by following some standard patterns.

—Rasul Shafikov,
Associate Instructor, M014: College Algebra, Indiana University

The key to success in a math course is to work regularly. As a rule, each new topic is based on the previously covered material. Failure to understand one topic may well result in poor understanding of the whole subject. The other important thing is to do a lot of problems. Practice makes perfect.

Try to work out all homework problems yourself. If there is something you cannot solve, it might be a good idea to discuss it with your classmates. If this doesn't clarify the matter, ask your instructor. But avoid asking questions without giving them a thought first. The instructor's answer will probably be forgotten if you don't spend some time trying to figure out the question yourself.

—Rasul Shafikov, Associate Instructor, M014: College Algebra, Indiana University

TIP: Most math instructors include variations on homework problems in their exams. In many cases, exam problems consist exclusively of such variations.

Many mathematics instructors advise students to get into the habit of translating diagrams as well as equations into words. For example, encountering arc-sin x, read "the arc sine of x" or "the inverse sine of x." Bring numerical expressions to the level of language and the level of language to numerical expressions.

Your instructor expects you to acquire facility working with diagrams as well as equations.

Great Expectations

Generally speaking, instructors and professors who teach introductory courses in various subjects expect great variation in the level of preparation among the students who come to them. Depending on your college or university, however, this may be somewhat less true among math instructors. Course placement is often at least partially determined by diagnostic tests, which tends to narrow the range of variation in any given class. If a math course is a core curriculum requirement at your school or is a prerequisite or requirement for many courses and programs, it is still probable that the instructor expects wide variation among students. If, however, most students who elect even an introductory math course intend to major in more or less math-intensive fields, the instructor most likely expects a reasonably sound high school level of preparation.

Avoid unpleasant surprises. If you have doubts about the level of your preparation, talk it over with a math advisor or instructor. Be honest and straightforward. Find out just what the instructor expects you to walk through the door with.

You should also discuss what is expected in class. While some instructors want you to memorize and master extensive amounts of prescribed materials, others are more interested in your learning basic principles, which you are then expected to apply to problems. Still others are most interested in your acquiring practical problem-solving skills, period.

It is the case, however, that, whatever else they look for, most math instructors expect you to rely on your own initiative. In any math class, it helps to be a self-starter rather than someone who expects to be spoon-fed the course contents.

You should know that, at minimum, your instructor expects you to attend class, take lecture notes, read the textbook, do all assigned homework problems, and do practice problems on your own. In addition, you are expected to:

◆ Think problems through

◆ Pay close attention to demonstration problems in class

◆ Study formulas, equations, diagrams, tables, and graphs as carefully as you study text

◆ Ask questions

The main course objective is to strengthen the students' algebra and problem-solving skills so they will be better prepared to fulfill the general college math requirement. Required skills include

◆ Simplifying and combining variable expressions
◆ Solving and graphing equations
◆ Working with functions
◆ Working with mathematical models (one-variable word problems)

The course examinations test these skills with problems that reflect the content and level of difficulty of the assigned text exercise. You should:

◆ Take thorough notes on the main ideas presented in each chapter covered, placing special emphasis on articulating key concepts "in their own words."
◆ Work practice problems and seek help when necessary.

The exams account for 100 percent of the final grade. To do well on them, you should:

◆ Study main ideas in the context of specific problems.
◆ Practice with "closed book, timed" practice tests.

—Mark A. McCombs, Director of Teacher Training, Math 10: College Algebra, University of North Carolina

Ask questions—but not right away. Don't ask to be handed the solution to a problem that puzzles you. Work on your own first to try to figure it out. A demanding math class presents you with an opportunity to assume a leadership role. Consider organizing a study group. Your initiative will be appreciated—by fellow students and the instructor alike.

Course objectives include teaching and/or strengthening the algebra skills needed for further math classes and for classes in other areas of study. Students should be able to solve problems of the various types seen in the exams. They should be able to factor most common polynomials and understand that algebraically solving problems may lead to solutions that do not make sense in the problem (negative length of sides of a polygon, square root of negative, or dividing by zero).

To succeed in this course, at the very minimum, students must complete all assignments. I would also suggest going through the problem sets in the book and practicing the problems that appear similar to the ones assigned.

Since this course is designed mainly to teach and reinforce problem-solving skills, it is practice and experience that will prove most beneficial.

—Ryan P. Albert, Graduate Teaching Assistant, Math 104: Introductory College Level Algebra, Ohio State University

CHAPTER 3

KEYS TO SUCCESSFUL STUDY

THIS CHAPTER OUTLINES THE GENERAL SKILLS THAT ARE INDISPENSABLE TO SUCCESSful study, with special emphasis on skills important to the study of mathematics.

But let's not start thinking about math just yet. Let's just start thinking. After all, isn't that what college and college courses are all about?

Well, not quite. *Think* about it.

For the ancient Greeks of Plato's day, about 428 to 348 BC, "higher education" really *was* all about thinking. Through dialogue, back and forth, teacher and student *thought* about physics, psychology, mathematics, the nature of reality—whatever. Perhaps the teacher evaluated the quality of the student's thought, but there is no evidence that Plato graded exams, let alone assigned students a final grade for the course.

Times have changed.

"Don't Study for the Grade!"

Today, you get graded. All the time, and on everything you do. Now, most of the professors, instructors, and teaching assistants from whom you take your courses will tell you that the "real value" of the course is in its contribution to your "liberal education." A professor may even solemnly protest that he or she does *not* "teach for" the midterm and final. Nevertheless, the introductory, survey, and core courses almost always *include* at least one midterm and a final, and these almost always account for a very large part of the course grade. Even if the professor can disclaim with a straight face *teaching for* these exams, few students would deny *learning for* them.

The truth is this: more than anything else, most curriculum core courses are focused on the midterm and final exams.

"Grades Aren't Important"

Let's keep thinking.

You, and most likely your family, are investing a great deal of time and cash and sweat in your college education. It would be pretty silly if the payoff for all those resources were a letter grade and a numerical GPA. Ultimately, of course, the benefit is knowledge, a feeling of achievement, an intellectually and spiritually enriched life— *and* preparation for a satisfying and (you probably hope) financially rewarding career.

But the fact is that if you don't perform well on midterms and finals, your path to all these forms of enrichment will be blocked. And the fact *also* is that your performance is measured by grades. Sure, almost any reason you can think of for investing in college is more important than amassing a collection of A's and B's, but those stupid little letters are part of what it takes to get you to those other, far more important, goals.

"Don't Study for the Exam"

Most professors hate exams and hate grades. They believe that prodding students to pass tests and then evaluating their performance with a number or letter makes the whole process of education seem pretty trivial. Those professors who tell you that they "don't teach for the exam" may also advise you not to "study for the exam." That's not exactly what they mean. They want you to study, but to study in order to *learn,* not to pass an exam.

It's well-meaning advice, and it's true that if you study for the exam, intending to ace it, then promptly forget everything you've "learned," you are making a pretty bad mistake. Yet these same professors are part of a system that demands exams and grades, and if you don't study for exams, the chances are very good that you won't make the grade and you won't achieve the higher goals you, your family, and your professors want for you.

Lose-Lose or Win-Win? Your Choice

When it comes to studying, especially in your introductory-level courses, you have some choices to make. You can decide grades are stupid, not study, and perform poorly on the exams. You can try not to study for the exams, but concentrate on higher goals, perform poorly on the exams, and never have the opportunity to reach those higher goals. You can study for the exams, ace them all, then flush the information from your memory banks, collect your A or B for the course, and move on without having learned a thing. These are all lose-lose scenarios, in which no one—neither you nor your teachers (nor your family, for that matter)—gets what he or she really wants.

Or you can go the win-win route.

We've used the words *goal* and *goals* several times. For an army general, winning the war is the goal, but to achieve this goal certain *objectives* must first be accomplished—such as winning battle number one, number two, number three, and so on. Objectives are intermediate goals or steps toward an ultimate goal.

Now, put exams in perspective. Performing well on an exam need not be an alternative to achieving higher goals, but should be an objective necessary to achieving those "higher goals."

The win-win scenario goes like this: use the fact of the exams as a way of focusing your study for the course. Focus on the exams as immediate objectives, crucial to achieving your ultimate goals. *Do* study for the exam, but not solely for the exam. *Don't* mistake the battle for the war, the objective for the goal; but *do* realize that you must attain the objectives in order to achieve the goal.

And that, ladies and gentlemen, is the purpose of this book:

◆ To help you ace the midterm and final exams in fundamental math courses . . .

◆ . . . without forgetting everything you learned after you've aced them.

This guide will help you use the exams to master the course material. It will help you make the grade—and actually learn something in the process.

Focus

How many times have you read a book word for word, finished it, and closed the cover—only to realize that you've learned almost nothing from it? Unfortunately, it's something we all experience. It's not that the material is too difficult or that it's over our heads. It's that we mistake reading for studying.

For many of us, reading is a passive process. We scan page after page, the words go in, and, alas, the words seem to go out. The time, of course, goes by. We've *read* the book, but we've *retained* all too little of what we've read.

Studying—even studying math—certainly involves reading, but reading and studying aren't one and the same activity. Or we might put it this way: studying is intensely focused reading.

How do you *focus* reading?

Begin by setting objectives. Now, saying that your objective in reading a certain number of chapters in a textbook or reading your lecture notes is to learn the material is not very useful. It is an obvious but vague *goal* rather than a well-defined *objective*.

Why not let the approaching exam determine your objective?

"I will read and retain the stuff in chapters 10 through 20 because that's what's going to be on the exam, and I want to ace the exam."

Now you at least have an objective. Accomplish this *objective*, and you will be on your way to achieving the *goal* of learning the material.

◆ An *objective* is an immediate target. A *goal* is for the long term.

Concentrate

To move from passive reading to active study requires, first of all, concentration. Setting up objectives (immediate targets) rather than looking toward goals (long-term targets) makes it much easier for you to concentrate. Few of us can (or would) put our personal lives completely on hold for four years of college, several years of graduate school, and an unknown number of years in the working world in order to concentrate on achieving a goal. But just about anyone can discipline him- or herself to set aside distractions for the time it takes to achieve the objective of studying for an exam.

Step 1. Find a quiet place to work.

Step 2. Clear your mind. Push everything aside for the few hours you spend each day studying.

Step 3. Don't daydream—*now*. Daydreaming—letting your imagination wander—is actually essential to real learning. But right now you have a specific objective to attain. This is not the time for daydreaming.

Step 4. Deal with your worries. Those pressing matters you can do something about *right now*, take care of. Those you can't do anything about, push aside for now.

Step 5. Don't worry about the exam. Take the exam seriously, but don't fret. Instead of worrying about the prospect of failure, use your time to eliminate failure as an option.

Plan

Let's go back to that general who knows the difference between objectives and goals. Chances are he or she also knows you'd better not march off to battle without a plan. Remember, you don't want to *read*. You want to *study*. This requires focusing your work with a plan.

The first item to plan is your time.

Step 1. Dedicate a notebook or organizer-type date book to the purpose of recording your scheduling information.

Step 2. Record the following:
   a. Class times
   b. Assignment due dates
   c. Exam dates
   d. Extracurricular commitments
   e. Study time

Step 3. Inventory your various tasks. What do you have to do today, this week, this month, this semester?

TIP: If you are in doubt about what tasks to assign the highest priority, it is generally best to allot the most time to the most complex and difficult tasks and to get these done first.

Step 4. Prioritize your tasks. Everybody seems to be grading you. Now's *your* chance to grade the things they give you to do. Label high-priority tasks A, middle-priority tasks B, and lower-priority tasks C. This will not only help you decide which things to do first, it will also aid you in deciding how much time to allot to each task.

Step 5. Enter your tasks in your scheduling notebook and assign order and duration to each according to its priority.

Step 6. Check off items as you complete them.

Step 7. Keep your scheduling book up to date. Reschedule whatever you do not complete.

Step 8. Don't be passive. Actively *monitor* your progress toward your objectives.

Step 9. Arrange and rearrange your schedule to get the most time when you need it most.

Packing Your Time

Once you have found as much time as you can, pack it as tightly as you can.

Step 1. Assemble your study materials. Be sure you have all necessary textbooks and notes on hand. If you need access to library reference materials, study in the library. If you need access to reference materials on the Internet, make sure you're at a computer.

Step 2. Eliminate or reduce distractions.

Step 3. Become an efficient reader and note taker.

An Efficient Reader

Step 3 requires further discussion. Let's begin with the way you read.

Nothing has greater impact on the effectiveness of your studying than the speed and comprehension with which you read. If this statement prompts you to throw up your hands and wail, "I'm just not a fast reader," don't despair. You can learn to read faster and more efficiently.

Consider taking a speed-reading course. Take one that your university offers or endorses. Most of the techniques taught in the major reading programs actually do work. Alternatively, do it yourself.

Step 1. When you sit down to read, try consciously to force your eyes to move across the page faster than normal.

Step 2. Always keep your eyes moving. Don't linger on any word.

Step 3. Take in as many words at a time as possible. Most slow readers aren't slow-witted. They've just been taught to read word by word. Fast and efficient readers, in contrast, have learned to read by taking in groups of words. Practice taking in blocks of words.

Step 4. Build on your skills. Each day, push yourself a little harder. Move your eyes across the page faster. Take in more words with each glance.

Step 5. Resist the strong temptation to fall back into your old habits. Keep pushing.

When you review material, consider skimming rather than reading. Hit the high points, lingering at places that give you trouble and skipping over the stuff you already know cold.

> **TIP:** Are you—ugh!—a *vocalizer*? A vocalizer is a reader who, during "silent" reading, either mouths the words or says them mentally. Vocalizing greatly slows reading, often reduces comprehension, and is just plain tiring. Work to overcome this habit—*except* when you are trying to memorize some specific piece of information. Many people do find it helpful to verbally repeat a sentence or two in order to memorize its content. Just bear in mind that this does not work for more than a sentence or paragraph of material.

An Interactive Reader

Early in this chapter we contrasted passive reading with active studying. A highly effective way to make the leap from passivity to activity is to become an *interactive* reader.

Step 1. Read with a pencil in your hand.

Step 2. Use your pencil to underline key concepts. Do this consistently. (That is, always read with a pencil in your hand.) Don't waste your time with a ruler; underscore freehand.

Step 3. Underline *only* the key concepts. If everything seems important to you, then up the ante and underline only the absolutely *most* important passages.

Step 4. If you prefer to highlight material with a transparent marker (a Hi-Liter, for example), fine. But you'll still need a pencil or pen nearby. Carry on a dialogue with your books by writing condensed notes in the margin.

Step 5. Put difficult concepts into your own words—right in the margin of the book. This is a great aid to understanding and memorization.

Step 6. Link one concept to another. If you read something that makes you think of something else related to it, make a note. The connection is almost certainly a valuable one.

Step 7. Comment on what you read.

TIP: The physical act of underlining actually helps you memorize material more effectively—though no one is quite sure why. Furthermore, underlining makes review skimming more efficient and effective.

TIP: Some instructors advise against using Hi-Liter style markers to underscore books and notes, because doing so may discourage you from writing notes in the margin. If holding a Hi-Liter means you won't also pick up a pencil to engage in a lively dialogue with books or notes, then it is best to lay aside the Hi-Liter and take the more actively interactive route with pencil alone.

TIP: Some students are reluctant to write in their textbooks, because it reduces resale value. True enough. But is it worth an extra five dollars at the end of the semester if you don't get the most out of your multi-thousand-dollar and multi-hundred-hour investment in the course?

TIP: Because math problems can be demanding and, at the very least, require fresh concentration, it is best not to set yourself the goal of finishing a *certain number of problems* in any given study period. Instead, set an *amount of time* for the study period and solve as many problems as you can during that time.

Problems, Problems

Algebra, trigonometry, and calculus are conceptual as well as problem-related disciplines. While reading quickly and efficiently is an important skill even in math courses, problem solving is even more critical. Problem solving involves:

Step 1: Understanding the underlying concept(s)

Step 2: Recognizing what approaches and equations to plug into the concept(s)

Step 3: Applying the approaches, theorems, principles, and equations

Step 4: Calculating to solve the equations

The first two steps require intelligent study of your text and intelligent attention to lectures. The last two steps require *practice*. Application and calculation become easier—more "automatic"—with practice. This is analogous to learning the basics of music. Music history and music theory require study, but learning to play an instrument requires, above all else, practice.

Do your homework problems regularly. Do extra problems for practice. You cannot learn to solve problems by reading about them or by reading about the mathematical principles related to them. You learn to solve problems by actually working them.

Math concepts at any level are challenging enough. You do *not* need the additional challenge of enrolling in a course for which your background has inadequately prepared you. Perhaps a diagnostic placement test is required for a given course—or perhaps one is not required, but is available. If you have any misgivings about the adequacy of your math background for a certain course, take the test. Beyond this:

◆ Talk to the instructor of the course you are considering.

◆ Consider working with a math tutor recommended by the instructor or the math department.

◆ Be sure to dust off what math skills you have. How? Practice!

Taking Notes

The techniques of underlining, highlighting, paraphrasing, linking, and commenting on textbook material can also be applied to your classroom and lecture notes.

Of course, this assumes that you have taken notes. There are some students who claim that it is easier for them to listen to a lecture if they avoid taking notes. For a small minority, this may be true; but the vast majority of students find that note taking is essential when it comes time to study for midterms and finals. This does not mean that you should be a stenographer or court reporter, taking down each and every word. To the extent that it is possible for you, absorb the lecture *in your mind*, then jot down major points, preferably in loose outline form.

Become sensitive to the major points of the lecture. Some lecturers will come right out and tell you, "The following three points are key." That's your cue to write them down. Other cues include:

◆ Repetition. If the lecturer repeats a point, write it down.

◆ Excitement. If the lecturer's voice picks up, if his or her face becomes suddenly animated, if, in other words, it is apparent that the material is of particular interest to the speaker at this point, your pencil should be in motion.

◆ Verbal cues. In addition to such verbal elbows in the rib as "this is important," most lecturers underscore transitions from one topic to another with phrases such as "Moving on to . . ." or "Next, we will . . ." or the like. This is your signal to write a new heading.

◆ Slowing down. If the lecturer gives deliberate verbal weight to a word, phrase, or passage, make a note of it.

◆ Visual aids. If the lecturer writes something on a blackboard or overhead projector or uses it in a computer-generated presentation, make a note.

◆ Equations. If the instructor presents an equation as key, believe it. Write it down carefully. (Most core-level math instructors provide key formulas and diagrams on handouts. Many even include equation and formula sheets with the exams they give. However, don't rely on having these aids available. Write down any formulas and equations presented in lecture.)

Filtering Notes

Some students take neat notes in outline form. Others take sprawling, scrawling notes that are almost impossible to read. Most students can profit from *filtering* the notes they take. Usually, this does *not* mean rewriting or retyping your notes. Many instructors agree that this is a waste of time. Instead, they advise, underscore the most important points, filtering out the excess.

Should you create an outline of your notes?

While it is probably helpful to outline major concepts, don't spend too much time creating an elaborate formal outline. Instead, invest that time merely underlining or highlighting the most important concepts.

Tape It?

Should you bring a tape recorder to class? The short answer is probably not. To begin with, some instructors object to having their lectures recorded. Even more important, however, is the tendency to complacency a tape recorder creates. You might feel that you don't have to listen very carefully to the live lecture, since you're getting it all on tape. This could be a mistake, since the live presentation is bound to make a greater impression on you, your mind, and your memory than a recorded replay.

So much for the majority view on tape recorders; however, a few professors actually *recommend* taping lectures. Fewer still suggest that students refrain from taking written notes in class, except for copying mathematical examples. These instructors recommend that you play back the tape later and take notes from that. If this method appeals to you, make sure of the following:

◆ Clear with your instructor the use of a tape recorder. Make sure he or she has no objections.

◆ Before you rely on the tape recorder, make sure that it works, that its microphone adequately picks up what's going on in the classroom, and that you are well positioned to record. The back of a large lecture hall is probably not a good place to be.

◆ Don't zone out or daydream during the lecture, figuring that you don't need to pay attention because you're getting it all on tape anyway. The value of the tape-recorded lecture is as a repetition and reinforcement of what you have already absorbed.

◆ Do copy down all mathematical examples—or obtain handouts on these.

◆ Make certain that you really do play back—and take notes from!—the taped lectures. This means setting aside the time to repeat the entire lecture.

Build Your Memory

Just as a variety of speed-reading courses are available, so a number of memory-improvement courses, audio tapes, and books are on the market. It might be worth your while to scope some of these out, especially for memory-intensive courses like math or if you are planning ultimately to go into a field that requires the memorization of a lot of facts. In the meantime, here are some suggestions for building your memory:

◆ Be aware that most so-called memory difficulties are really learning difficulties. You "forget" things you never learned in the first place. This is usually a result of passive reading or passive attendance at lectures—the familiar in-one-ear-and-out-the-other syndrome.

◆ Memorization is made easier by two seemingly opposite processes. First, we tend to remember information for which we have a context. It will indeed be hard to remember a bunch of definitions if you try to study these out of context—as a list rather than as parts of an interrelated system of thought.

◆ Second, memorization is often made easier if we break down complex material into a set of key phrases, words, or concepts.

◆ It follows, then, that the best way to build your memory where a certain subject is concerned is to try to understand information in context. Get the big picture.

◆ It also follows that, even if you have the big picture, you may want to break down key concepts into a few key words or phrases.

How We Forget

It is always better to keep up with classwork and study than it is to fall behind and desperately struggle to catch up. This said, it is nevertheless true that most forgetting occurs within the first few days of exposure to new material. That is, if you "learn" 100 facts about Subject A on December 1, you may forget 20 of those facts by December 5 and another 10 by December 10, but by March 1 you may still remember 50 facts. The curve of your forgetting tends to flatten. Eventually, you may forget all 100 facts, but you will forget fewer and fewer each week.

Now, what does this mean to you?

It means that, midway through the course, you need to review material you learned earlier. You cannot depend on having mastered it forever just because you studied it two, three, four, or more weeks earlier.

REMEMBER: The mere fact that you have tape recorded a lecture does not mean that you have learned the material. Technology is indeed marvelous, but it won't perform miracles. You have to listen to the tape and take notes from it.

TIP: Memorization is important, make no mistake, especially in the information-rich mathematical disciplines. However, brute memory *is* usually overrated. Virtually all of the instructors and professors who have contributed to this book counsel students to *think* rather than merely memorize.

TIP: Many memory experts suggest that you try to put the key terms you identify into some sort of sentence, then memorize the sentence. Others suggest creating an acronym out of the initials of the key words or concepts. No one who lived through the Watergate scandal in the 1970s can forget that President Nixon's political campaign was run by CREEP (Committee to *RE*-*E*lect the *P*resident), and long after everything else taught in high school biology is forgotten, many students remember the sentence they memorized in order to learn how biologists classify organisms: *K*en *P*ut *C*andy *O*n *F*red's *G*reen *S*ofa (kingdom, phylum, class, order, family, genus, species). In trig, many find it helpful to remember the processes for solving general triangles by categorizing them with labels according to what is known about the triangle: SSS is associated with the procedures to be used when three sides are known, SAS is used when two sides and the included angle are known, ASA when two angles and the included side are given, AAS is used for two angles and a nonincluded side, and SSA for two sides and a nonincluded angle. The point: use whatever memory aids help *you*.

The Virtues of Cramming

Ask any college instructor about last-minute cramming for an exam, and you'll almost certainly get a knee-jerk condemnation of the practice. But maybe it's time to think beyond that knee jerk.

TIP: Any full-time college student studies several subjects each semester. This makes you vulnerable to *interference*—the possibility that learning material from one subject will interfere with learning material in another subject. Interference is usually at its worst when you are studying two similar or related subjects. If possible, arrange your study time so that work on similar subjects is separated by work on an unrelated one.

Let's get one thing absolutely straight. You cannot expect to pack a semester's worth of studying into a single all-nighter. It just isn't going to happen, especially in a math class! However, cramming can be a valuable supplement to a semester of conscientious studying.

◆ You forget the most within the first few days of studying. (Or have you forgotten?)

Well, if you cram the night before the exam, those "first few days" won't fall between your studying and the exam, will they?

◆ Cramming creates a sense of urgency. It brings you face to face and toe to toe with your objective. Urgency concentrates the mind.

◆ Assuming you aren't totally exhausted, material you study within a few hours of going to bed at night is more readily retained than material studied earlier in the day.

Burning the midnight oil may not be such a bad idea.

Cramming Cautions

Then again, staying up late before a big exam may not be such a hot idea, either. Don't do it if you have an early-morning exam. And don't transform cramming into an all-nighter. You almost certainly need *some* sleep to perform competently on tomorrow's exam.

Remember, too, that while cramming creates a sense of urgency, which may stimulate and energize your study efforts, it may also create a feeling of panic, and panic is never helpful.

Cramming is *not* a substitute for diligent study and regular practice throughout the semester. But just because you have studied and practiced diligently, don't shun cramming as a supplement to regular study, a valuable means of refreshing the mind and memory.

Polly Want a Cracker?

We've been talking a lot about memory and memorization. It's an important subject and, for just about any course of study, an important skill. Some subjects—the mathematical disciplines included—are more fact and memory intensive than others.

However, beware of relying too much on simple, brute memory. Try to assess what the professor really wants: students who demonstrate on exams that they have absorbed the facts the instructor and the textbooks have dished out? Or students who demonstrate such skills as critical thinking, synthesis, analysis, and imagination?

Most math instructors want students to be familiar with essential theorems, formulas, and procedures, but they also value evidence of *understanding* that goes deeper than rote memorization.

> **TIP:** If you hate cramming, don't do it. It's not for you, and it will probably only raise your anxiety level. Get some sleep instead.

Use This Book (and Get Old Exams)

One way to judge what the professor values and expects is to pay careful attention in class. Is discussion invited? Or does the course go by the book and by the lecture? Also valuable are previous exams. Many professors keep these on file and allow students to browse through them freely. Fraternities, sororities, and formal as well as informal study groups sometimes maintain such files, too. These days, previous exams may even be posted on the university department's World Wide Web site. Most math professors encourage students to pore over past exams and mine them for practice problems.

Of course, you are holding in your hand a book chock full of sample and model midterm and final exams. Read them. Study them. And let them focus your study and review of the course.

Study Groups: The Pros and the Cons

In the old days (whenever that was), it was believed that teachers taught and students learned. More recently, educators have begun to wonder whether it is possible to *teach* at all. A student, they say, learns by teaching him- or herself. The so-called teacher (who might be better termed a "learning facilitator") helps the student teach him- or herself.

Maybe this is all a matter of semantics. Is there really a difference between *teaching* and *facilitating learning*? And between *learning* and *teaching oneself*? The more important point is that the focus in education has turned away from the teacher to the student, and students in turn have often responded by organizing study groups in which they help each other study and learn.

These can be very useful:

◆ In the so-called "real" world (that is, the world after college), most problems are solved by teams rather than individuals.

◆ Many people come to an understanding of a subject through dialogue and question and answer.

◆ Studying in a group (or even with one partner) makes it possible to drill and quiz one another.

◆ In a group, complex subjects can be broken up and divided among members of the group. Each one becomes a "specialist" on some aspect of the subject, then shares his or her knowledge with the others.

◆ Studying in a group may improve concentration.

◆ Studying in a group may reduce anxiety.

Not that study groups are without their drawbacks:

◆ All too often, study groups become social gatherings, full of distraction (however pleasant) rather than study. This is their greatest pitfall.

◆ All members of the group must be highly motivated to study. If not, the group will become a distraction rather than an aid to study, and it is also likely that friction will develop among the members, some of whom will feel burdened by "freeloaders."

◆ The members of the study group must not only be committed to study, but to one another. Study groups fall apart—bitterly—if members, out of a sense of competition, begin to withhold information from one another. This must be a Three Musketeers deal—all for one and one for all—or it is worse than useless.

◆ The group may promote excellence—or it may agree on mediocrity. If the latter occurs, the group will become destructive.

In summary, study groups tend to bring out the members' best as well as worst study habits. It takes individual and collective discipline to remain focused on the task at hand, to remain committed and helpful to one another, to insist that everyone shoulder his or her fair share, and to insist on excellence of achievement as the only acceptable standard—or at least the only valid reason for continuing the study group.

CHAPTER 4

SECRETS OF SUCCESSFUL TEST TAKING

THIS CHAPTER PRESENTS GENERAL ADVICE FOR SUCCESSFUL TEST TAKING. THE NEXT chapter offers some strategies specific to taking tests in math.

Sometimes it seems that the difference between academic success and something less than success is not smarts versus non-smarts or even study versus non-study, but simply whether or not a person is good at taking tests. That phrase—"good at taking tests"—was probably first heard back when the University of Bologna opened for business late in the eleventh century. The problem with phrases like this is that they are true enough, yet not very helpful.

Fact: Some people are and some people are not "good at taking tests."

So what? Even if successful test-taking doesn't come to you easily or naturally, you *can* improve your test-taking skills. Now, if you happen to have a knack for taking tests, well, congratulations! But that won't help you much if you neglect the kind of preparation discussed in the previous chapter.

Why Failure?

In analyzing performance on most tasks, it is generally better to begin by asking what you can do to succeed. But, in the case of taking tests, success is largely a question of avoiding failure. So let's begin there.

When the celebrated bank robber Willy "The Actor" Sutton was caught, a reporter asked him why he robbed banks. "Because that's where they keep the money," the handcuffed thief replied. At least one answer to the question of why some students perform poorly on exams is just as simple: "Because they don't know the answers."

There is no magic bullet in test taking. But the closest thing to one is knowing the course material cold. Pay attention, keep up with reading and other assignments, attend class, listen in class, take effective notes—in short, follow the recommendations of the previous chapter—and you will have taken the most important step toward exam success.

Have you ever gotten your graded exam back, with disappointing results, read it over, and found question after question you realized you could have answered correctly?

"I *knew* that!" you exclaim, smacking yourself in the forehead.

What happened? You really did prepare. You really did know the material. What went wrong?

TIP: Most midterm and final exams really *are* representative of the course. If you have mastered the course material, you will almost certainly be prepared to perform well on the examination. Very few instructors purposely create deceptive exams or trick questions, or even questions that require you to think beyond the course. Most instructors are interested in creating exams that help you and them evaluate your level of understanding of the course material. Think of the exams as natural, logical features of the course, not as sadistic assignments designed to trip you up. Remember, your success on an examination is also a measure of the instructor's success in presenting complex information. Very few teachers can—or want to—build careers on trying to fail their students.

Anxiety Good and Bad

The great American philosopher and psychologist William James (1842–1910) once advised his Harvard students that an "ounce of good nervous tone in an examination is worth many pounds of . . . study." By "good nervous tone" James meant something very like anxiety. You should not expect to feel relaxed just before or during an exam. Anxiety is natural.

Anxiety is "natural," and it can be helpful. "Good nervous tone," alert senses, sharpened perception, and adrenalin-fueled readiness for action are natural and healthy responses to demanding or threatening situations. We are animals, and these are reactions we share with other animals. The mongoose that relaxes when it confronts a cobra is a dead mongoose. The student who takes it easy during the math final . . . well, the point is neither to fight anxiety nor to fear it. Accept it, and even welcome it as an ally. Unlike our hominid ancestors of distant prehistory, we no longer need the biological equipment of anxiety to help us fight or fly from the snapping saber teeth of some animal of prey, but every day we do face challenges to our success. Midterms and finals are just such challenges, and the anxiety they provoke is real, natural, and unavoidable. It may even help us excel.

What good can anxiety do?

◆ Anxiety can focus our concentration. It can keep the mind from wandering. This makes thought easier, faster, and, often, more acute and effective.

◆ Anxiety can energize us. We've all heard stories about a 105-pound mother who is able to lift the wreckage of an automobile to free her trapped child. This isn't fantasy. It really happens. And just as adrenalin can provide the strength we need when we need it most, it can enhance our ability to think under pressure.

◆ Anxiety moves us along. Anxious, we work faster than when we are relaxed. This is valuable since, in most midterms and finals, limited time is part of the test.

◆ Anxiety prompts us to take risks. We've all been in classes in which the instructor has a terrible time trying to get students to speak and discuss and venture an opinion. "Come on, come on," the poor prof protests, "this is like pulling teeth!" Yet, when exam time comes, all heads are bent over blue books or answer sheets, and the answers—*some* kind of answers—are flowing forth or, at least, grinding out. Why? Because the anxiety of the exam situation overpowers the inertia that keeps most of us silent most of the time. We take the risks we have to take. We answer the questions.

◆ Anxiety can make us more creative. This is related to risk taking. Necessity, the old saying goes, is the mother of invention. Phrased another way: *we do what we have to do*. Under pressure, many students find themselves taking fresh and creative approaches to problems.

So don't shun anxiety. However, realize that, unfortunately, the scoop on anxiety isn't all good news, either.

Anxiety evolved as a mechanism of *physical* survival. Biologists and psychologists talk of the fight or flight response. Anxiety prepares a threatened animal either to fight the threat or to flee from it. The action is physical and, typically, very short term. In our "civilized" age, the threats are generally less physical than intellectual and emotional, and they tend to be of longer duration than a physical fight or a physical flight. This means that the anxiety mechanism does not always work to enhance our chances for survival, or at least our chances to survive the course by performing well on exams. Some of us are better than others at adapting the *physical* benefits of anxiety to the *intellectual* and *emotional* challenges of an exam. Some of us, unfortunately, are unable to benefit from anxiety, and, for still others of us, anxiety is downright harmful. Here are some of the negative effects anxiety may have on exam performance:

◆ Anxiety can make it difficult to concentrate. True, anxiety focuses concentration. But if it focuses concentration on the anxious feelings themselves, you will have less focus left over for the exam. Similarly, anxiety may cause you to focus unduly on the perceived consequences of failure.

◆ Anxiety causes carelessness. If anxiety can prompt you to take creative risks, it can also cause you to rush through material and, therefore, to make careless mistakes or simply to fail to think through a problem or question.

◆ Anxiety distorts focus. Anxiety may impede your judgment, causing you to give disproportionate weight to relatively unimportant matters. For example, you may become fixated on solving a lesser problem at the expense of a more important one. This is related to the next point.

◆ Anxiety may distort your perception of time. You may think you have more or less of it than you really do. The result may be too much time spent on a minor question at the expense of a major one.

◆ Anxiety tends to be cumulative. Many test takers have trouble with a question early in the exam, then devote the rest of the exam to worrying about it instead of concentrating on the rest of the exam.

> **TIP:** Exam questions are battles in a war. No general expects to win every battle. Accept your losses and move on. Dwell on your losses and you will continue to lose. Note that in most math problems partial credit is given for whatever you do manage to accomplish, especially if you are careful to show your work.

◆ Anxiety drains energy. For short periods of time, it can be energizing and invigorating. But if anxiety becomes chronic, it begins to tire you out. You do not perform as well.

◆ Anxiety can keep you from getting the rest you need. If it is generally unwise to stay up all night *studying* for an exam, how much less wise is it to stay up uselessly *worrying* about one?

How can you combat anxiety?

Step 1: Don't fight it. Accept it. Remember, anxiety is a natural response to a stressful situation. Remember, too, that some degree of anxiety aids performance. Try to learn to accept anxiety and *use* it. Let it sharpen your wits and stoke the fires of your creativity.

Step 2: Don't worry about how you feel. Focus on the task. Usually, you will feel better once you overcome the initial jitters and inertia. William James, who lauded "good nervous tone," also once observed that we do not run because we are frightened, but we are frightened because we run. If you concentrate on your fear and act as if you are afraid, you will become even more fearful.

Step 3: Prepare for the exam. Do whatever you must to master the material. Build confidence in your understanding of the course, and your anxiety should be reduced.

Step 4: Get a good night's sleep before the exam.

Step 5: Avoid coffee and other stimulants. Caffeine tends to increase anxiety. (However, if you are a caffeine fiend, don't pick the day or two before a big exam to kick the habit. You *will* suffer withdrawal symptoms.)

Step 6: Try to get fresh air shortly before the exam. This is especially valuable if you have been cooped up for a long period of study. Take a walk. Get a look at the wider world for a few minutes.

Have a Plan, Make a Plan

A large component of destructive anxiety—probably the largest—is fear of the unknown. Reduce anxiety by taking steps to reduce the component of the unknown.

Step 1: To repeat—do whatever is necessary to master the material on which you will be examined.

Step 2: Use the exams in this book to familiarize yourself with the kinds of exams you are likely to encounter.

Step 3: If possible, study old exams actually given in the course.

Let's pause here before going on to Step 4. Just reading over the exams in this book or leafing through exams formerly given in the course will not help you much.

◆ Analyze the types of questions asked and problems presented. Are they concept oriented, or do they stress demonstration of calculating skills? Do the problems call for reliance on rote memory and the application of memorized formulas, or are they more "think" oriented, requiring significant initiative to answer?

◆ Don't just predict which questions you could or could not answer or which problems you could or could not solve. Try actually answering some of the questions and solving some of the problems.

◆ Consider taking a full-scale practice exam. If you do so, be sure to time yourself.

◆ If you are looking at sample exams with solutions, evaluate the solutions. How would you grade them? What would you do better?

◆ Don't just sit there, *do* something. If your analysis of the sample exams or old exams reveals areas in which you are weak, address those weaknesses.

An effective way to reduce the unknown is to create a plan for confronting it. Let's go on to Step 4.

Step 4: Make a plan. Begin well before the exam day. Decide what areas you need to study hardest. Based on your textbook notes and—especially—on your lecture notes, try to anticipate what kinds of problems will be on the exam. Work up strategies and solutions for these problem types.

Step 5: Make sure you've done the simple things. The night before your exam, make certain you have whatever equipment you'll need. If you will be allowed to use reference materials, bring them. If you are permitted to use a calculator, make certain your batteries are fresh. Make certain you have plenty of pencils and paper. Bring a watch.

TIP: More serious stimulant drugs ("speed," "uppers") are *never* a good idea. They are both illegal and dangerous (possibly even deadly) and, for that matter, their effect on exam performance is unpredictable. The chances are they will impede performance rather than aid it (though you may erroneously *feel* that you are doing well). Also avoid over-the-counter stimulants. These are caffeine pills and will probably increase anxiety rather than improve performance.

TIP: Don't make the mistake of devoting all of your time to trying to make last-minute repairs to weak spots ("I've got one hour to read that textbook I should have been reading all along!") only to ignore your strengths. Develop your strengths. With any luck at all, the exam will give you an opportunity to show yourself at your best—not just trip you up at your worst. Be as prepared as you can be, but remember, there is nothing wrong with excelling in a particular area. Play to your strengths, not your weaknesses.

Step 6: Expect a shock. The first sight of the exam usually packs a jolt. At first sight, questions may draw a blank from you. Problems you were sure would appear on the exam will be absent, and some you never expected will be staring you in the face. Don't panic. *Everybody feels this way.*

TIP: Do plenty of practice problems with two objectives in mind: First, simply to get better and more efficient at working the problems; second, to uncover and diagnose your weak spots. Discover the gaps you need to fill—the kind of problems you need to work harder on.

Step 7: Write nothing yet. Read through the exam thoroughly. Be certain that you (a) understand any instructions and (b) understand the problems. This is especially important with word problems, which are all too easy to misinterpret.

Step 8: If you are given a choice of which problems to work on, choose them now. Unless the items vary in the point value assigned to them, choose those that you feel most confident about solving. Don't challenge yourself.

Alternative Step 8: If you are required to address all the test items, identify those about which you feel most confident. Answer these first.

Step 9: After you have surveyed the exam, create a "time budget": note—jot down—how much time you should give to each major question or problem.

Step 10: Reread the question or problem before you begin to write. Then plan your answer or solution.

TIP: When you study for an exam, it is usually best to assign high priority to the most complex and difficult issues, devote ample time to these, and master them first. When you take an exam, however, and you are under time pressure, tackle first what you can most readily and thoroughly answer, then go on to more doubtful tasks. Your professor will be more favorably impressed by good solutions than by failed attempts to solve problems you find difficult. (*Do show your work!* Even failed solutions typically get partial credit for the work that is done correctly.)

Plan Your Solution

Here is where practice really helps! If playing the piano required relearning scales and chords each time a musician put fingers to keys, there would be very few pianists in the world. But, of course, experienced pianists approach their instrument with a well-rehearsed knowledge of scales and chords. With this background, built by practice, even a brand-new score looks at least somewhat familiar.

If you have read your text, listened in lecture, and, most of all, practiced problem solving, the problems you see on the exam will likewise look familiar. Indeed, most math professors purposely put modified versions of homework problems on their exams.

Begin, then, by surveying the problems. Identify those that are most familiar. Work on those first—and use on them the procedures that you have practiced.

Some problems are relatively easy. Some are tough. Many problems, however, are not so much difficult as complex. That is, they are made up of a number of subproblems. Identify the subproblems and subroutines required to solve them. Identify and observe the order in which the subproblems should be solved. This is a very important step in planning your solution.

Plan Your Answer

While most math exams (including those in this book) are heavily weighted toward problem solving, some introductory courses also call for answering conceptual or information-related questions.

Perhaps you have heard a teacher or professor comment on the exam he or she has just handed out: "The answers are in the questions." This kind of remark is more helpful than it may seem at first.

Begin by looking for the key words in the question. These are the verbs that tell you what to do, and they typically include:

> **TIP:** Reduce the number of unknowns you carry into the exam. Practice the basic mathematical operations you know you will need. Be very certain you are fully familiar with the advanced functions of your calculator. You'll have your hands full doing the math that appears on the exam. You can't afford to take time to learn the math you should have at your fingertips.

◆ Compare

◆ Contrast

◆ Criticize

◆ Define

◆ Describe

◆ Discuss

◆ Evaluate

◆ Explain

◆ Illustrate

◆ Interpret

◆ Justify

◆ Outline

◆ Relate

◆ Review

◆ State

◆ Summarize

◆ Trace

Of these key words, the following are most often found in the kind of short-answer questions you are likely to encounter in the conceptual portion of a math exam:

◆ To *define* something is to state the precise meaning of the word, phrase, or concept. Be succinct and clear.

◆ To *illustrate* is to provide a specific, concrete example.

◆ To *outline* is to provide the main features or general principles of a subject. This need not be in paragraph or essay form. Often, outline format is expected.

◆ To *state* is similar to *define*, though a statement may be even briefer and usually involves delivering up something that has been committed to memory.

◆ To *summarize* is to state briefly—in sentence form— the major points of an argument or position or concept, the principal features of an event, or the main events of a period.

Approaching the Short-Answer Test

The short-answer exams you may get in math courses are of two major kinds:

1. *Recall exams* include questions that call for a single short answer (usually there is a single correct answer), and fill-in-the-blank questions in which you are asked to supply missing information in a statement or sentence.

2. *Recognition exams* include multiple-choice tests, true-false tests, and matching tests (matching one item from column A with one from column B).

If the exam is a long one and time is short, invest a few minutes in surveying the questions, so that you can be certain to answer those you are confident of, even if they come near the end of the exam.

Be prepared to answer multiple-choice questions through a process of elimination if necessary. Usually, even if you are uncertain of the one correct answer among a choice of five, you will be able to eliminate one, two, or three answers you know are incorrect. This at least increases your odds of giving a correct response.

Unless your instructor has specifically informed you that he or she is penalizing guesses (actually taking points away for incorrect responses versus awarding zero points to unanswered questions), *do* guess the answers even to those questions that leave you in the dark.

Plan your responses to true-false questions carefully. Look for telltale qualifying words and phrases, such as *all, always, never, no, none,* or *in all cases.* Questions with such absolute qualifiers often require an answer of *false,* since relatively few general statements are always either true or false. Conversely, questions containing such qualifiers as *sometimes, usually, often,* and the like are frequently answered correctly with a response of *true.*

A final word on guessing: first guess, best guess. Statistical evidence consistently shows that a first guess is more likely to be right than a later one. Obviously, if you have responded one way to a question and then the correct answer suddenly dawns on you, *do* change your response. But if you can choose only from a variety of guesses, go with your first or "gut" response.

CAUTION! Many math courses do include multiple-choice exams—that is, exams in which the right answer is included among choices presented to you. However, in many cases, providing the answer is not enough to earn full credit for the item. If you are required to show your work—to show how you arrived at the answer—be sure to do so. Obviously, having to show your work makes guessing a less viable option.

Take Your Time

Yes, yes, yes, this is easier said than done. But the point is this:

◆ Plan your time.

◆ Work efficiently, but not in a panic.

◆ Make certain your responses are legible. This is especially important in calculations and equations.

◆ Take time to double-check your calculations.

◆ In word responses, take time to spell correctly. Even if an instructor does not consciously deduct points for misspelling, such basic errors will negatively influence his or her evaluation of the exam.

◆ In responses that require a sketch or a graph, make sure what you draw is legible, and make doubly sure you have labeled all features correctly and clearly.

◆ If a short answer is called for, make it short. Don't ramble.

◆ If instructed to show your work, make certain it is presented clearly and legibly.

◆ Use *all* of the time allowed. The instructor will not be impressed by a demonstration that you have finished early. If you have extra time, reread the exam. Look for careless errors. Do not, however, heap new guesses on top of old ones; where you have guessed, stick with your first guess.

CHAPTER 5

MATH EXAM STRATEGIES

THE PRECEDING CHAPTER PROVIDED SOME GENERAL ADVICE ON TAKING TESTS. THIS chapter presents some test-taking strategies that are specific to math exams.

Study

Begin by accepting this sobering fact: no test-taking strategy will substitute for knowing the material. Study and practice throughout the semester. Use the approaching exams to focus your study. Intensify your study efforts in the days before a major examination.

When you study, simply keep the idea of an examination in the back of your mind. In studying, try to work practice problems in the same way you will work problems on the exam. That is:

◆ When you start to study a new textbook chapter or topic, begin by surveying (skimming) the material to get a general idea of the topic. This is also what you should do in approaching the exam. Begin by surveying the whole exam.

◆ Read all instructions, explanations, and specifications before working a problem.

◆ When you work practice problems, concentrate on the *why* and the *how*. The *why* is the reasoning that governs the problem. The *how* is the method or process used to solve the problem. Concentrate on these now, so that they will become (to the extent possible) second nature by the time of the exam.

◆ In studying textbook problems that have been worked out, pay special attention to the order or sequence of steps used in the solution. Develop a sense of order and the habit of solving problems in the proper sequence of steps. This will save a great deal of time during the exam, and it will reduce anxiety profoundly.

◆ When you solve a practice problem, test yourself thoroughly. That is, don't be content with simply arriving at a solution. Consider why you did each step. Be certain you can explain, in words, your reasons for solving the problem the way you did.

◆ Don't let mistakes throw you. Learn from them.

What All Problems Have in Common

It's all too easy to let a difficult problem bury you. Don't let this happen. Go into the exam prepared. Not only should you study and practice, but you should make it your business to learn everything you can about what kinds of problems will be on the exam. In most cases, this is hardly a state secret:

◆ Your instructor will probably tell you what to expect.

◆ If your instructor doesn't tell you, ask. (It couldn't hurt!)

◆ Look at exams given previously.

◆ Review key homework problems. These are the source of most exam problems.

◆ Review examples of problem types covered in lecture.

Whatever else you can determine about what will be on the exam, know that each problem will definitely involve answering three basic questions:

1. What is given? What are the facts in the problem?

2. What is unknown? What is the problem asking for? What is the answer to be found?

3. How do I proceed? What formulas, methods, and steps are required to solve this problem?

Strategies for Word Problems

Many exam questions in algebra, as well as trig and calculus, are word problems. That is, you are asked to derive certain information from a set of data given. Word problems are valuable because they link mathematical processes with real-world

It is important to
1. **Do the homework each day as soon as possible after class. Before reading the theory, read the questions for focus.**
2. **Talk with other students first about anything that is unclear; mathematics gains immensely from being verbalized, and the value of collaboration cannot be overemphasized. Then, if necessary, ask a tutor (a free service provided at our university, staffed by a graduate or undergraduate student) or the instructor for the solutions. I post the solutions in the library as well.**

I give a short list of review problems prior to each exam (about a week before, so that the students can still ask questions); each question on the exam is exemplified by a set of review problems (taken from the homework). Then, I recommend that a student give himself or herself a "practice exam," based on those review problems, especially timing the performance. The most successful class I ever held was one to which I was able to give such a practice test myself (attendance was totally optional). Unfortunately, there is not enough time for that to be common practice.

—Emma Previato, Professor, MA 123: Calculus I, Boston University

After each topic is covered in class, work problems from that section until they seem easy. If problems are still a bit difficult, this probably means that you have not yet mastered the material. You may have to do more problems than those that were assigned, but you'll be much more prepared.

Before each test, be sure to work review problems. Being an active studier is the key here. Many students think that if they skim over the book and their old homework, they have studied. In order to retain material, you must approach it actively. If you want to skim over the book, for example, you'll probably find it very helpful to jot down key points, or sketch an outline as you go.

If you begin to get frustrated and are spending what you think is an excessively long time working problems, seek help and advice! Don't wait until you're having serious problems. See your instructor when you have questions. The instructor will be able to give you an idea of about how long your homework should be taking to complete, and give you study tips specifically for their class.

—Stephanie L. Fitch, Lecturer,
Math 6: Trigonometry,
University of Missouri-Rolla

applications; however, they often pose difficulties beyond mastery of concepts or the complexity of computation. The trouble begins with confusion over what is given and what is being asked.

Adopt a Systematic Solving Technique

Begin by identifying what is being asked. Underline it, if necessary. This is a crucial first step.

Having identified what is being asked, you should now be able to pull out the information you are given. Isolate the givens from the wording of the problem, Circling or underlining this information helps. If you can, translate the given information into a picture.

Now, using the information—what is being asked for and what is given—try to set up an equation or other straightforward system to solve the problem.

Use your equation or system to solve the problem.

◆ Compute carefully.

◆ Make absolutely certain that you are working in the same units. Word problems may include a mixture of units—for example, pounds and ounces or inches and feet. Convert the units into one common unit. Failure to take this elementary step is a common source of error.

Check to ensure that you have answered the question. Students commonly misinterpret the real question in a word problem. Answer what is being asked. Then check your answer. The most basic way of checking is just to use common sense. Is your answer reasonable or unreasonable?

Key Words in Word Problems

Even complex word problems include very common mathematical operations. Look for the following key words in word problems:

Added to
Addition
Increase
More than

Plus
Sum
Total

All of these tell you to *add*. The following tell you to *subtract*:

Decrease(d)

Word problems are notorious for providing more information than is necessary to arrive at a solution. Identify the information you need and discard the rest.

Difference

Fewer

Less

Less than

Minus

Reduced

And these call for *multiplication*:

At (for example: "the cost of 15 and a half yards of cotton *at* 55 cents per yard")

Of (for example "a third *of* the group")

Product ("the *product* of 4 and 3 is")

Times ("there are three *times* as many nonsmokers as smokers")

Total (may be an indicator of multiplication—for example: "If you spend $80 a week on groceries, what is the *total* for a three-month period?")

Twice (as in "*twice* the value of some number")

Finally, *division* is signaled by *quotient* and *ratio*, in addition to the obvious *divided into* or *divided by*. Also look for phrases such as "*half* the people who attended," in which the word *half* tells you to divide by 2.

Some Common Word Problem Scenarios

In algebra, common word problems relate to

◆ Compound interest

◆ Ratio and proportion

◆ Number problems

◆ Motion problems

◆ Number problems involving two variables

Compound Interest

Consider this problem:

Joe invests $100 at a 15 percent annual interest rate compounded yearly. What will be his final total after four years?

Begin your solution by underlining what is being asked for: the *final total after four years*. Now note what is given: $100, the 15 percent rate, and the fact that the interest is compounded yearly. That the interest is compounded yearly for four years tells us we must arrive at a solution in four parts.

Reduce the word problem to an equation:

Interest = principal x *rate* x *time*

$$I = prt$$
$$I = 100(.15)1$$
$$I = 115$$

Take the product, 115, and repeat the equation for the next year. Then take that product and solve for year three. Finally, take the year three product and solve for year four.

Ratio and Proportion

A worker can turn out 500 widgets in 5 hours. How many hours will it take for this worker to turn out 750 widgets, assuming he or she works at the same rate?

Again, begin by identifying what is asked for: *how many hours*. Next, identify the categories given: *hours* and *widgets*. Use the categories to set up an equation:

$$\frac{widgets}{hours} = \frac{widgets}{hours}$$

Then just plug in the numbers:

$$\frac{500}{5} = \frac{750}{x}$$

Cross-multiply:
$$500x = 5(750)$$
$$500x = 3750$$

$$\frac{500x}{500} = \frac{3750}{500}$$

$$x = 7.5 \text{ hours}$$

Number Problems

Problems involving numbers nested in words are quite common on exams. Here's a simple one.

When six times a number is increased by four, the result is forty. What is the number?

What do you have to find? The *number*. So set up an equation in which x stands for the number and that incorporates the other givens:

$6x + 4 = 40$

Solve the equation, subtracting 4 from each side:

$6x = 36$

Divide by 6:

$x = 6.$

Motion Problems

These occur frequently as algebra word problems and can be quite complex. Here's a simple one for the sake of illustration:

A car travels at 65 mph. How long will it take to travel 15 miles?

The answer wanted is obvious: *how long will it take?* Now you need to know the simple formula for computing distance: *distance = rate × time.* Plug in the givens and solve:

$d = rt$

$15 = 65(t)$

$15/65 = 65t/65$

$.23 = t$

It will take .23 hour to travel 15 miles.

Number Problems Involving Two Variables

Here's an example:

The sum of two numbers is 15. The difference between these two numbers is 7. Find the two numbers.

Begin as usual, by identifying what is being sought: *the two numbers.* You'll need to set up an equation with two unknowns. You've been given the sum of these unknowns as well as the difference. "Sum" tells you to add:

$x + y = 15$

"Difference" tells you to subtract:

$x - y = 7$

Add the equations:

$$\begin{array}{rcl} x \;+\; y &=& 15 \\ x \;-\; y &=& 7 \\ \hline 2x &=& 22 \\ \text{So } x &=& 11. \end{array}$$

Next, plug the result into the first equation:

$$\begin{array}{rcl} 11 \;+\; y &=& 15 \\ \text{So } y &=& 4 \end{array}$$

The numbers are 11 and 4.

Simplify by Plugging in Numbers

Just as word problems are simplified by reducing them to the underlying equations and numbers involved, so problems involving variables may be made simpler by plugging in numbers. Here is a typical multiple-choice question:

If x is a positive integer in the equation $12x = q$, then q is
  a. A negative odd integer
  b. A positive odd integer
  c. A negative even integer
  d. A positive even integer
  e. Zero

Plugging in numbers turns this abstract problem into a matter of simple arithmetic. It is given that x is a positive integer, so let's start by plugging in the simplest positive integer:

$$12(1) = q$$
$$12 = q$$

Plug in the next obvious choice:

$$12(2) = q$$
$$24 = q$$

Try another number. Let's do 3:

$$12(3) = q$$
$$36 = q$$

Any positive integer plugged into x yields a positive even integer. So the correct answer is d.

Three Multiple-Choice Strategies

In many introductory algebra courses, multiple-choice test formats are used, at least for a portion of the exam. In cases where you are not required to show your work, guessing becomes a reasonable last-ditch option if you really get stuck. However, blind guessing usually is neither necessary nor advisable.

You can simply try eliminating unreasonable choices. That will improve your odds of hitting on the correct answer.

Even better is working from the given answers. For example:

Find the square root of 2025:
    a. 15
    b. 25
    c. 35
    d. 45
    e. 55

Square roots can be tough without a calculator. But since the square root asks the question, *What number multiplied by itself equals the square* (in this case, 2025)?, you can simply multiply each of the given numbers by itself. Start in the middle, with 35 times 35. This way you'll know whether to move up or down on your next try. Since the product of 35 times itself is 1225, we know we have to move up, not down. Try 45 times 45, and you get 2025. So the answer is d.

Finally, when you are presented with a difficult calculation, try rounding off each of the multiple-choice options and working with those simplified numbers in order to approximate the correct answer.

One Picture Is Worth a Thousand Numbers

You have one more strategy in your arsenal. Be certain to use any diagrams provided with a question. If a diagram isn't provided, draw one. The more complex the problem, the more necessary a diagram probably is. Be sure to mark diagrams clearly for your use, including all the given facts.

A Word About Calculators

Make certain you understand the instructor's policy on the use of calculators in class, at home, and during exams. Use the calculator as a tool, not a crutch. And, because the calculator *is* a tool, learn to use it properly and efficiently. Practice with it. Become thoroughly familiar with all of its capabilities before using it in an exam. Exam day is not the day to start exploring the wonders of your calculator!

PART TWO

STUDY GUIDE

CHAPTER 6

PRE-ALGEBRA BACKGROUND

THE CHAPTERS THAT FOLLOW IN THIS PART OF *ACE YOUR MIDTERMS AND FINALS: Fundamentals of Mathematics* are designed as outlines of what you may expect to encounter in introductory algebra, trigonometry, and calculus courses. They are brief overviews and are not intended as math reviews or condensed textbook chapters. Think of them as menus that list the kinds of information, concepts, and problems you'll be dealing with in the introductory courses and, therefore, on midterms and finals. (You'll find a list of comprehensive study guides and reviews in Chapter 24, "Suggested Reading.")

This book does not include exams for "remedial" math courses: pre-algebra courses that generally cannot be counted toward core curriculum requirements but that help students fill in gaps left by spotty high school preparation. However, it is useful to begin with a survey of the kind and level of pre-algebra competency expected of students enrolling in a basic college algebra course.

Knowledge of Basic Terminology and Conventions

Chapters 23 and 24 provide, respectively, a glossary of essential math terms and a list of common math symbols. A few of the most important terms you'll need to know as you start an algebra course include:

Integers: These are the numbers . . . −2, −1, 0, 1, 2, . . .

Rational numbers: Fractions that can be written in the form *a/b*, with *a* an integer and *b* a *natural number* (one of the "counting numbers," 1, 2, 3, 4, . . .). Note that whole numbers may be rational numbers, because they can be written in fractional form: 6/1 is the number 6.

Irrational numbers: These cannot be written in the form *a/b* (*a* an integer and *b* a natural number). The square root of 3 and π are examples of irrational numbers.

Prime numbers: A prime has exactly two factors or can be evenly divided by only itself and 1. The number 5 is an example of a prime. Note that 2 is the only even prime number.

Composite numbers: In contrast to a prime, a composite is divisible by more than just 1 and itself. All even numbers except 2 are composites. An example of an odd composite is 9.

Squares: The result of a number multiplied by itself.

Cubes: The result when a number is multiplied by itself twice:
$2 \times 2 \times 2 = 8$. Thus $2^3 = 8$.

Basic Mathematical Operations

Pre-algebraic knowledge should encompass understanding of the basic operations of addition, subtraction, multiplication, and division.

Important properties of addition include:

◆ *Closure:* All answers fall into the original set. If two even numbers are added, the answer is even, and therefore the set of even numbers has closure under addition. The sum of two odd numbers is even, and therefore the set of odd numbers has no closure.

◆ *Commutative:* Addition is commutative, meaning that the order of the numbers added makes no difference. (This is not the case with subtraction.)

◆ *Associative:* Not only does order make no difference in addition, neither does grouping of the numbers. Thus addition, in contrast to subtraction, is associative.

◆ Any number added to 0 gives the original number. Therefore, 0 is the *identity element* of addition. Also, any number plus its *additive inverse* equals the identity, 0: $-4 + 4 = 0$.

Multiplication shares some of these basic properties with addition:

◆ Multiplication of two even numbers yields an even product. Thus, for multiplication, the set of even numbers has closure. In contrast to addition of odd numbers, however, the multiplication of odds yields an odd product, so the set of odd numbers also has closure under multiplication.

◆ As with addition, multiplication is *commutative*: the order in which numbers are multiplied makes no difference. This is not the case with division.

◆ The *identity element* for multiplication is not 0 but 1. The *multiplicative inverse* is the *reciprocal* of the number; that is, any number multiplied by its reciprocal equals 1 ($4 \times 1/4 = 1$).

Multiplication by 0 always equals 0. Zero divided by any number always equals 0; however, no number can be divided by 0. This is "undefined" and not permitted.

Powers

In the number 2^3, the superscript 3 (above and to the right of the quantity) is an *exponent*, which expresses the *power* to which the quantity is raised ($2 \times 2 \times 2 = 8$). An exponent may be negative, lowering the quantity by the expressed power.

The *square* of a number is that number raised to the power of 2—that is, multiplied by itself. The *cube* of a number is the number raised to the power of 3—that is, multiplied twice by itself.

You will need to come into algebra understanding how to do operations with powers and exponents.

The *square root* of a number n is the number y that, multiplied by itself (y^2), equals n. The *cube root* of a number n is the number y that, multiplied by itself twice (y^3), equals n.

> Note that working with roots and *radicals* (the square, cube, or other root of a number) is common in algebra.

Grouping Numbers

Pre-algebra knowledge should include familiarity with using parentheses () to group numbers or variables. Operations within parentheses must be performed before other operations. Brackets [] may be used to group numbers and variables that include groups in parentheses. Braces {} group numbers and variables that include both parentheses and brackets.

Grouping brings up the subject of *orders of operations*. It is important to perform operations in the following order:

1. Parenthetical operations

2. Powers and square roots

3. Multiplication or division (whichever comes first, left to right)

4. Addition or subtraction (whichever comes first, left to right)

Signed Numbers

Numbers to the right of 0 on a number line are positive. Those to the left are negative. You will work extensively with positive and negative numbers, including fractions. Know how to add, subtract, multiply, and divide signed whole numbers and signed fractions, and how to simplify fractions.

In some equation formats, bigger parentheses (instead of brackets) are used to group numbers and variables that include groupings in smaller parentheses.

Decimals, Percentages, and Scientific Notation

Pre-algebra competency includes working with decimals and the ability to convert fractions into decimals. You should also understand that a fraction whose denominator is 100 is a percent.

Mathematicians and scientists write very large numbers in *scientific notation* to avoid awkwardness. A number written in scientific notation is any number between 1 and 10 multiplied by a power of 10. Thus $3.4 \times 10^6 = 3,400,000$.

It's easy to convert scientific notation to common form: just move the decimal point to the right by the number of the exponent. You should know how to multiply and divide numbers expressed in scientific notation.

CHAPTER 7

ALGEBRA: THE BASIC CONCEPTS

S IMPLY STATED, ALGEBRA IS ARITHMETIC GENERALIZED OR ABSTRACTED SO THAT some of the numbers are replaced by letters (*variables*). As professional mathematicians see it, algebra is a broader and more complex subject than this simple definition suggests. *Elementary algebra*—basic algebra or college algebra, as represented in this book—is indeed calculating with variables instead of just numbers, and it includes solving equations. Advanced algebra, or *higher algebra,* is the study of abstract mathematical structures in which there are operations that have the properties of addition and multiplication. At the introductory level, we are dealing with elementary algebra.

What is the significance of algebra's abstraction? Any statement made in algebra is true for all numbers, and not just specific cases.

Set Theory

Basic algebra often begins with a review of *set theory,* which can provide a basis for precisely defining higher concepts and for mathematical reasoning. A *set* is a group of numbers, variables, or even objects. A member of a set is referred to as an *element.* Note the following:

- ◆ $2 \in \{1,2,3\}$ is read, "Two is an element of the set of 1,2,3."
- ◆ $\{2,3\} \subset \{1,2,3\}$ is read, "The set 2,3 is a subset of the set 1,2,3.
- ◆ $\varnothing$ and $\{\}$ are read as "the null set" or "the empty set."

Sets may be described by *rule* or by *roster*. *Rule* describes the elements of a set, whereas *roster* lists them. Sets can also be represented pictorially with *Venn diagrams* and *Euler circles*.

Describing sets by rule uses its own system of notation. For example, N is "the set of natural numbers and 0"; N* is "the set of natural numbers." An asterisk denotes the exclusion of 0 from a set. Other symbols include:

Z: the set of all integers

Z^+: the set of positive integers

Z^-: the set of negative integers

Z*: the set of all integers except 0

Q: the set of all rational numbers

Q^+: the set of positive rational numbers

Q^-: the set of negative rational numbers

Q*: the set of all rational numbers except 0

R: the set of all real numbers

R^+: the set of positive real numbers

R^-: the set of negative real numbers

R*: the set of all real numbers except 0

C: the set of all complex numbers

C*: the set of all complex numbers except 0

Operations with sets include *union*, in which two or more sets form a new set containing all the members of both or all sets, and *intersection*, in which two or more sets intersect to form a new set containing only the members of both or every set.

Union is expressed: $\{1,2,3\} \cup \{3,4,5\} = \{1,2,3,4,5\}$

Intersection is expressed: $\{1,2,3\} \cap \{3,4,5\} = \{3\}$

Sets present a graphical method of manipulating numbers as well as variables, illustrating possible relationships among them and fundamental laws governing such properties as (for example) commutation and association—concepts mentioned in the previous chapter. For this reason set theory can be an excellent entry into algebra.

Algebraic Expressions

An *algebraic expression* is any mathematical expression in which letters, called *variables*, are used to represent numbers. A *variable* is a letter or other symbol used to represent a number.

Variables and/or numbers may be combined by multiplication or division to create *terms*. An example of a term 5*xy*, in which 5 and the variables *x* and *y* are combined by multiplication; *x*/5 is a term in which 5 and *x* are combined by division.

Terms that are combined by addition, subtraction, or both are called *expressions*.

Algebra is a symbolic language in which algebraic expressions, letters, and variables are effectively substituted for words and numbers. The English-language question "What number increased by 5 gives 10 as a result?" may be expressed algebraically as $x + 5 = 10$.

When algebra is applied to specific disciplines, engineering, say, or business—it is important to ensure that the variables are defined. The familiar economics formula, "cost equals the selling price less the margin of profit" may be expressed in the formula $C = S - M$. To ensure that the formula is intelligible, it should be defined:

$C = S - M$, in which C is cost, S is the selling price, and M is the margin of profit.

Working with Expressions
Expressions can be added, subtracted, multiplied, and divided.

Addition and Subtraction
The addition and subtraction of expressions are among the most basic algebraic operations you will use. Numbers represented by the same variables can be added and subtracted; those containing unlike symbols cannot. Unlike terms must be expressed separately. Also, when combining signed numbers, expressions with the same signs are combined first.

Multiplication and Division
Multiplication is very basic to algebra and can be indicated, as it is in conventional arithmetic, with a times sign, x; with a centered dot, •; with parentheses, $(a)(b)$; or by mere juxtaposition, *ab* (this is called *understood multiplication*). Division may be indicated by a division sign, ÷, or slash, /: 4/2 is 4 divided by 2.

The *law of exponents* states that to multiply powers of the same base, add their exponents, so that (for example):
$$(3^5)(3^3) = 3^8$$

Conversely, to divide powers of the same base, subtract the exponent of the divisor from that of the dividend.
$$3^5/3^3 = 3^2$$

Evaluating Expressions

To *evaluate an expression* is to replace the unknowns with grouping symbols, insert the values for the unknowns, and do the arithmetic. For example:

$xy + z$ if $x = 2$, $y = 3$, and $z = 4$

$2(3) + 4 =$

$6 + 4 = 10$

CHAPTER 8

ALGEBRA: EQUATIONS

I N A MATHEMATICAL STATEMENT, NUMBERS AND VARIABLES ARE ROUGHLY ANALOGOUS to words; terms and expressions are roughly analogous to phrases; and *equations* correspond to complete sentences. An equation is a statement that shows, with numbers and variables, that two mathematical expressions are equal. The numerical values of variables that make the equation true are the *roots* or *solutions* of the equation. The solution *satisfies the equation.*

Broadly speaking, equations are of two types:

◆ *Identities*: An identity is an equation that holds true for all values of the variable.

◆ *Conditional equations*: A conditional equation is true only for certain values of the variables. These are the types of equations you will work with most frequently.

The concept of the equation is one of the most profound of human inventions and a tool basic not only to mathematics, but to civilization itself. Although the Greek mathematician-philosophers used equations between 540 and 250 BC, mathematical statements approaching the equation are found much earlier, in the Rhind Papyrus, an Egyptian document dating to 1650 BC (and believed to be based on an even earlier source).

Note that algebraic expressions consisting of a single term are *monomial*. Examples are $5x$, $2x^3$, $6xyz^2$. *Polynomials* consist of two or more terms, as in $x + y + z$. A *binomial* has exactly two terms, and a *trinomial* exactly three. Both of these, of course, are polynomials.

Essential Axioms

As the sentence is basic to communication in words, so the equation is basic to mathematical communication. Working with equations is the major focus of introductory algebra.

Underpinning equations are the *axioms of equality*:

◆ Reflexive axiom: $a = a$

◆ Symmetric axiom: If $a = b$, then $b = a$

◆ Transitive axiom: If $a = b$ and $b = c$, then $a = c$

◆ Additive axiom: If $a = b$ and $c = d$, then $a + c = b + d$

◆ Multiplicative axiom: If $a = b$ and $c = d$, then $ac = bd$

The Equation as Balance

As you advance in algebra during the course of even a single semester—and perhaps move on to higher branches of mathematics later—equations can become highly complex, as is apparent from the sample exams included in this book. Nevertheless, no matter how complex the equation, it works like a balance, with the equal sign functioning as the fulcrum or balance point. To balance an equation, do the same thing to both sides of the equal sign. Here is a simple example:

$$x - 10 = 5$$

Get x by itself on one side by adding 10 to both sides:

$$\begin{aligned} x - 10 &= 5 \\ +10 \ &+10 \\ \hline x \quad\ &= 15 \end{aligned}$$

Equations do not require numbers. In algebra, you will often work with *literal equations*, which contain only letters. They are governed by the same axioms and principles as equations with numbers. For example:

$AB - C = D$
Solve for A

$$AB - C = D$$
$$\underline{+\,C \qquad +\,C}$$
$$AB \qquad = D + C$$

$$\frac{AB}{B} = \frac{D + C}{B}$$
$$A = \frac{D + C}{B}$$

Ratios and Proportions

A *ratio* compares two or more numbers or variables and, in an equation, appears in fractional form as *a/b* or $\frac{a}{b}$. A ratio is read as "*a* is to *b*." *Proportions* are written as two ratios equal to each other:

$$\frac{a}{b} = \frac{c}{d}$$

Many algebra problems involve ratios. For example:

Solve for x

$$\frac{4}{x} = \frac{2}{5}$$

$$(4)(5) = 2x$$
$$20 = 2x$$

$$\frac{20}{2} = \frac{2x}{2}$$
$$10 = x$$

More Complex Equations

We have illustrated some basic principles of equations by way of defining a subject that is central to any basic algebra course. Even basic college algebra moves quickly beyond these simple equations, however. Often, the next step in complexity is solving *systems of equations*, also known as *simultaneous equations*.

The simplest system involves two equations with the same two unknowns in each:

$$3x - 6y = 10$$
$$9x + 15y = -14$$

You will also work with systems of three equations in three unknowns, such as:

$$2x + 3y - 4z = 1$$
$$3x - y - 2z = 4$$
$$4x - 7y - 6z = -7$$

Two simultaneous equations are commonly solved by one of three methods:

1. Addition/subtraction

2. Substitution

3. Graphing

Introductory algebra courses cover all three methods extensively.

Addition/subtraction involves four major steps:

1. Multiplication of one or both equations by some number to make the numbers in front of one of the unknowns the same in each equation.

2. Addition or subtraction of the two equations to eliminate one unknown.

3. Solving for the other unknown.

4. Insertion of the value of the first unknown in one of the original equations to solve for the second unknown.

In the case of a three-equation system with three unknowns, you will learn how to derive from it a system of two equations with two unknowns, then work with these.

Substitution is a method that involves replacing one equation with another to arrive at a solution. For example, in the system

$$x = y + 8$$
$$x + 3y = 48$$

you could begin your solution by substituting $(y + 8)$ from the first equation for the x in the second:

$$(y + 8) + 3y = 48$$

Graphing, a method you will see in a number of the sample exams in this book, puts each equation on a *coordinate graph*. Graphing is a basic method of *analytic geometry* as well as trigonometry and calculus, but it is also a subject that is introduced in basic college algebra. We will take a look at coordinate graphing in the next chapter.

Quadratic Equations

The equations we've briefly discussed so far are *linear equations*, that is, equations involving only expressions whose variables are of degree 1, meaning that the powers of an expression add up to the first degree. (In contrast, the expression $3x^2$ has degree 2; $2x^2y^3$ has degree 5. Note that $4xy$ has degree 2, since the sum of the powers of x and y is 2.) Linear equations are "linear" because they are represented graphically as a straight line.

Quadratic equations are more complex than linear equations and involve an expression or expressions, containing a single variable, of degree 2. The form is this:

$$ax^2 + bx + c = 0$$

The quadratic term here is ax^2, because it is of degree 2. Note that while bx is also of degree 2, it does not contain a *single variable* of power 2; therefore, it is not a quadratic term, but a *linear term*.

There are three basic methods for solving quadratic equations:

1. Factoring

2. Using the quadratic formula

3. Completing the square

Factoring is generally the first approach to the quadratic that is learned in algebra. To *factor* is to find two or more quantities whose product equals the original quantity. There are five steps to solving a quadratic equation by factoring:

1. Put all terms on one side of the equal sign, so that only 0 is on the other side.

2. Factor.

3. Set each equal factor to 0.

4. Solve the resulting equations.

5. Check the solution by inserting the answer into the original equation.

Factoring cannot solve all quadratic equations. In cases where the roots (solutions) are not rational numbers, factoring will not work. A general solution of quadratic equations is provided by the quadratic formula:

$$x = \frac{-b \pm \sqrt{b^2 - 4ac}}{2a}$$

The variables a, b, and c are taken from the quadratic equation as written in the general form $ax^2 + bx + c = 0$ where a is the numeral that goes in front of x^2, b is the numeral inserted in front of x, and c is the numeral with no variable next to it.

Note that a part of the quadratic formula, the *radicand*, $b^2 - 4ac$, is called the *discriminant* (D) and is important because it determines the nature of the roots of the equation:

If $D > 0$, then the equation has two real, distinct roots.

If $D = 0$, then the equation has two real, duplicate roots.

If $D < 0$, then the equation has no real root.

Finally, quadratic equations may be solved by a method called *completing the square*. Begin by putting the equation in the form of $ax^2 + bx = -c$. You must ensure that $a = 1$. If it does not, multiply through the equation by $1/a$. Next, using the value of b from the new equation, add $(b/2)^2$ to both sides of the equation. This forms a perfect square on the left side of the equation. Next, find the square roots of both sides of the equation, then solve the equation that results.

CHAPTER 9

ALGEBRA: INEQUALITIES, GRAPHING, AND FUNCTIONS

WHEREAS AN EQUATION IS A STATEMENT OF EQUALITY, AN *INEQUALITY* IS A statement in which the relationships are not equal. Formally, instead of an equal sign, inequalities contain the symbols > (greater than), < (less than), ≥ (greater than or equal to), or ≤ (less than or equal to). As with equations, inequalities may be *conditional* or *unconditional*.

◆ Conditional inequalities are true only for certain values of the variables, as in $(x + 2) > 6$.

◆ Unconditional inequalities (also called absolute inequalities) are true for all values of the variables, as in $(x^2 + 1) > 0$.

As algebraic equations are governed by certain axioms and properties, so are inequalities:

◆ The *trichotomy axiom* is $a > b$, $a = b$, or $a < b$. That is, these are the only relationships possible between two numbers.

◆ The *transitive axiom* states: if $a > b$, and $b > c$, then $a > c$. Conversely, if $a < b$, and $b < c$, then $a < c$.

◆ The *addition property* of inequalities is defined in this way: if $a > b$, then $a + c > b + c$.

◆ The *positive multiplication property*: if $c > 0$, then $a > b$ only if $ac > bc$.

◆ The *negative multiplication property*: if $c < 0$, then $a > b$ only if $ac < bc$.

Solving Inequalities

Inequalities are solved in much the same way as equations, except that, if you multiply or divide both sides by a negative number, you have to reverse the direction of the sign. For example:

Solve for x:

$$-7x > 14$$

$$\frac{-7x}{-7} < \frac{14}{-7}$$

$$x < -2$$

Note that, because we divide both sides of the equation by -7, we must reverse the sign of the answer from $+2$ to -2.

Inequalities that include nonlinear members must be solved by factoring. For example:

$$x^2 - 3x - 10 > 0$$

is factored to

$$(x + 2)(x - 5) > 0$$

Note that the numerical values of the factors must have the same sign; they must be both positive or both negative. Two cases must be looked at:

$$\left.\begin{array}{l} x + 2 > 0 \mid x > -2 \\ x - 5 > 0 \mid x > 5 \end{array}\right\} x_1 > 5$$

$$\left.\begin{array}{l} x + 2 < 0 \mid x < -2 \\ x - 5 < 0 \mid x < 5 \end{array}\right\} x_2 < -2$$

If we start with

$$x^2 - 3x - 10 < 0$$

the values of the two factors must be opposite. We factor the equation to:

$$(x + 2)(x - 5) < 0$$

Then:

$$\left.\begin{array}{l} x + 2 > 0 \mid x > -2 \\ x - 5 < 0 \mid x < 5 \end{array}\right\} -2 < x < 5$$

$$\left.\begin{array}{l} x + 2 < 0 \mid x < -2 \\ x - 5 > 0 \mid x > 5 \end{array}\right\} \text{impossible}$$

Graphing

Integers and real numbers can be represented graphically on a *number line*, which is useful for visualizing inequalities and other relationships. The number line is also a good way to understand the concept of *absolute value*, which is the measure of a number's distance from 0 on the number line; hence, the absolute value of a number is either positive or 0. It cannot be negative. The number 5 has an absolute value of 5, and so does –5. Both are five steps away from 0:

$$|5| = 5$$
$$|-5| = 5$$

By adding to the horizontal number line a vertical line, we create the *Cartesian coordinate system*, which is used in analytic geometry as the basis for graphing lines on a plane and in algebra as the basis for graphing equations.

The basic coordinates look like this:

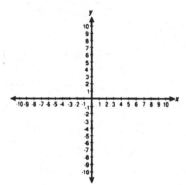

As each point on a number line is assigned a number, so each point on a plane is assigned a pair of numbers. For example, (+4,–3) would be indicated by a point placed at the intersection of a line drawn down from +4 on the horizontal line and another line drawn to the right from –3 on the vertical line. Such numbers are *ordered pairs*.

◆ The first number in an ordered pair is the *x-coordinate*.

◆ The second number in an ordered pair is the *y-coordinate*.

The coordinate graph is divided into quadrants I, II, III, and IV.

◆ Quadrant I is the upper right quarter of the graph. I, *x* is always positive, and I, *y* is always positive.

◆ Quadrant II is the upper left quarter of the graph. II, *x* is always negative, and II, *y* is always positive.

◆ Quadrant III is the lower left of the graph. III, *x* is always negative, and III, *y* is always negative.

◆ Quadrant IV is the lower right of the graph. IV, x is always positive, and IV, y is always negative.

Graphing Equations

One of the simplest graphing exercises is graphing a linear equation on the coordinate plane. This is done by finding the solutions to the equation by giving a value to one variable and solving the equation that results for the value of the other variable. The process is repeated with different values. Here is a straightforward example:

$$x + y = 4$$
$$(0) + y = 4$$
$$y = 4$$

$$(1) + y = 4$$
$$\underline{-1 \qquad = -1}$$
$$y = 3$$

$$(2) + y = 4$$
$$\underline{-2 \qquad -2}$$
$$y = 2$$

Draw up a chart of your results:

x	y
0	4
1	3
2	2

You now have a set of ordered pairs: (0,4), (1,3), and (2,2). It is a simple matter to plot these. This line (even though it is straight) is called a *curve*.

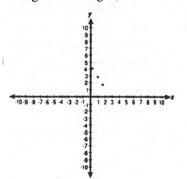

Draw a straight line connecting the dots, and you have graphed solutions for the equation.

Basic algebra at the college level typically involves a good deal of graphing, including the graphing of nonlinear equations. The curves of such equations truly are "curved" in the common sense; that is, the graphs are nonlinear.

Functions

A *function* is an association between two or more variables in which to every value of each of the *independent variables* (also called *arguments*) one value of the *dependent variable* corresponds. Here is a more concrete example. Say a pound of sugar costs $2. Two pounds of sugar will cost $4. A half pound of sugar will cost $1. The price of sugar is a *function* of weight.

Functions can be understood graphically. Here is the graph of the simple linear equation $y = x + 1$.

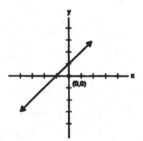

All functions are relations, but not all relations are functions. A *relation* consists of any set of ordered pairs. On the coordinate graph, the set of all the x's of a relation is called the *domain* of the relation. The set of all the y's is the *range* of the relation. A function is a relation in which each member of the domain is paired with exactly one element of the range. In the case of the equation graphed, $y = x + 1$, the domain and range of the function are both the set of real numbers, and the relation is, in fact, a function because for any value of x there is a unique value of y.

Not only do algebra problems often involve determining domain and range and whether a relation is a function, they may also call for finding the value of a function, which is the equivalent of finding the value of the range of the relation. Thus, given the function $f = \{(1,-3)(2,4)(-1,5)(3,-2)\}$, the value of the function at 1 is -3, at 2 it is 4, and so on. Expressed in conventional form, this would be:

$$f(1) = -3,$$
$$f(2) = 4$$

and so on. Read the equation "f of $1 = -3$" and interpret it as saying "the value of the function f at 1 is -3."

Variations

A *variation* is a relation between a set of values of one variable and a set of values of other variables. The graph of a *direct variation* passes through the *origin* [that is, the intersection of the x- and y-coordinates, (0,0)]. The function $y = kx$ is a direct variation, with k the *constant of proportionality* or the *constant of variation*.

A direct variation can also be written as a proportion, "y_1 is to x_1 as y_2 is to x_2":

$$\frac{y_1}{x_1} = \frac{y_2}{x_2}$$

x_1 and y_2 are the *means*, and y_1 and x_2 are the *extremes*. The product of the means equals the product of the extremes. Solve the proportion by multiplying the means and extremes, then proceed with the solution as usual.

Inverse variation, also called *indirect variation*, is expressed as $y = k/x$. As y increases, x decreases. The relationship is one of *inverse proportion*.

A Parting Word on Introductory Algebra

The material we have briefly surveyed in these last three chapters (plus, to some extent, Chapter 6) is basic and common to all introductory-level college algebra courses. Be aware, however, that what we've outlined is a departure point only. College-level courses develop these basic themes much more fully than is sketched here, both conceptually and in terms of the complexity of calculations and graphing involved. Moreover, many courses introduce additional concepts and techniques. The best way to gauge the scope—and intellectual demands—of the typical college algebra course is to review the sample exams you'll find in Chapters 12 through 17 of this book.

CHAPTER 10

TRIGONOMETRY OVERVIEW

TRIGONOMETRY IS THE BRANCH OF MATHEMATICS CONCERNED WITH FUNCTIONS OF angles and their applications to calculations in geometry. In many applications, the essential problem is the solution of triangles. Trigonometry enables us to find the unknown parts of triangles by arithmetical processes. Trigonometry is one of the most important branches of applied mathematics, since it is used extensively in engineering, surveying, cartography, navigation, and astronomy. Indeed, it was astronomy, beginning in the fifteenth century, that provided much of the impetus for the development of trigonometry as an independent branch of mathematics.

Trigonometry itself has two main branches. What has just been described—and is the subject of introductory-level trigonometry courses—is *plane trigonometry*, which is concerned with angles and distances in one plane. Application to similar problems in more than one plane is the province of *spherical trigonometry*.

Who Takes Introductory Trigonometry?

Among nonmath majors, students who have advanced beyond elementary algebra are likely candidates for an introductory-level trigonometry course. The course is also required as a prerequisite or adjunct to a variety of scientific and engineering fields. Trigonometry is also seen as excellent preparation for calculus.

Trigonometric Functions

Trigonometry begins with a consideration of simple right triangles. Because any triangle can be broken down into two right triangles, it is ultimately possible to use trigonometry to work with all kinds of triangles.

Geometry tells us that, in a right triangle BAC, $\angle A + \angle B = 90$. Also, we know that $c^2 = a^2 + b^2$.

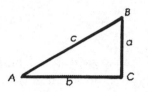

From the first equation, we can find one of the acute angles if the other is given. From the second equation, we can find the length of any side if the other two are given. But geometry does not provide a way to find $\angle A$ if given the two sides a and b. Trigonometry provides a method for calculating the angles if we know the sides, and for calculating the sides if we know the angles.

Functions of Acute Angles

Euclidian geometry states that if two angles of one triangle have the same measure as two angles of another triangle, then the two triangles are similar. Also, in similar triangles, the ratio of any two sides of one triangle equals the ratio of corresponding sides of the second triangle. If the angles of a triangle remain the same, but the sides increase or decrease, the ratios of the sides remain the same. Thus trigonometric ratios in right triangles are functions of the size of the angles. This is the cornerstone of trigonometry.

The key ratios are defined using a triangle inscribed within a circle, with the circle's radius as the triangle's hypotenuse. The equation is $x^2 + y^2 = r^2$, where r is the hypotenuse (radius), x is the side adjacent, and y is the side opposite.

$$\sin \alpha = \frac{y}{r} = \frac{\text{length of side opposite } \alpha}{\text{length of hypotenuse}}$$

$$\cos \alpha = \frac{x}{r} = \frac{\text{length of side adjacent to } \alpha}{\text{length of hypotenuse}}$$

$$\tan \alpha = \frac{y}{x} = \frac{\text{length of side opposite } \alpha}{\text{length of side adjacent to } \alpha}$$

The reciprocals of sine, cosine, and tangent are, respectively, the functions cosecant, secant, and cotangent. They have the following ratios:

$$\csc \alpha = \frac{r}{y}$$

$$\sec \alpha = \frac{r}{x}$$

$$\cot \alpha = \frac{x}{y}$$

If trigonometric functions of an angle θ are combined in an equation, and the equation is valid for all values of θ, the equation is a *trigonometric identity*. Key trigonometric identities are:

$$\frac{\sin \theta}{\cos \theta} = \tan \theta$$

$$\frac{\cos \theta}{\sin \theta} = \cot \theta$$

$$\sin^2 \theta + \cos^2 \theta = 1$$

Three pairs of trigonometric functions are *cofunctions*:

◆ Sine and cosine

◆ Tangent and cotangent

◆ Secant and cosecant

From the right triangle XYZ, these identities can be derived:

$$\sin X = \frac{x}{z} = \cos Y \qquad \sin Y = \frac{y}{z} = \cos X$$

$$\tan X = \frac{x}{y} = \cot Y \qquad \tan Y = \frac{y}{x} = \cot X$$

$$\sec X = \frac{z}{y} = \csc Y \qquad \sec Y = \frac{z}{x} = \csc X$$

So:

$$\sin \alpha = \cos (90° - \alpha) \qquad \cos \alpha = \sin (90° - \alpha)$$
$$\tan \alpha = \cot (90° - \alpha) \qquad \cot \alpha = \tan (90° - \alpha)$$
$$\sec \alpha = \csc (90° - \alpha) \qquad \csc \alpha = \sec (90° - \alpha)$$

Note that acute angles in standard position (on a *coordinate graph*; see Chapter 9) are all in the first quadrant; therefore, all of their trigonometric functions exist and are positive in value. This is not necessarily the case with angles in other quadrants.

Solving Triangles

Much of your early work in an introductory trigonometry course will involve solving triangles, beginning with right triangles. A triangle involves six basic measurements: the angles A, B, and C and the sides *a*, *b*, and *c*. If any three of these six measurements are known, the other values can be calculated. In a right triangle, of course, the measurement of one angle is always given as 90°.

Triangles other than right triangles are called *arbitrary triangles*.

◆ The *Law of Sines* is used to solve an arbitrary triangle when we know two angles and one opposite side or two sides and an opposite angle. The general formula, developed by Ptolemy of Alexandria about AD 150, is:

$$\frac{\sin A}{a} = \frac{\sin B}{b} = \frac{\sin C}{c}$$

◆ The *Law of Cosines* is used to solve an arbitrary triangle when two sides and the included angle are known or when three sides are known:

$$a^2 = b^2 + c^2 - 2bc \cos A$$
$$b^2 = a^2 + c^2 - 2bc \cos B$$
$$c^2 = a^2 + b^2 - 2bc \cos C$$

◆ The *Law of Tangents* is used to solve an arbitrary triangle if two sides and the included angle are known. Its general form was established in 1583 by the British mathematician Thomas Fincke:

$$\frac{a-b}{a+b} = \frac{\tan\frac{1}{2}(A-B)}{\tan\frac{1}{2}(A+B)}$$

Area of Triangles

The most familiar formula for finding the area of a triangle is $K = bh$, where K is the area of the triangle, b is the triangle's base, and h is its height. This formula can be combined with others to find the area of a triangle when two sides and the included angle are given or when two angles and a side are known.

When three sides are known, *Heron's formula* is used. The *semiperimeter* is found by $s = (a + b + c)$, then:

$$K = \sqrt{s(s - a)(s - b)(s - c)}$$

Trigonometric Identities

Equations that express relations among trigonometic functions that are true for all values of the variables involved are called *trigonometric identities.* Not only are such identities important in solving trig problems, they are powerful tools in calculus.

Trigonometric Equations

In trigonometric equations, the unknown quantity appears as a trigonometric function of one or more variables. If only one kind of function is present, the equation is *basic* and can be solved algebraically. If the equation contains several functions of one or several angles, it must be reduced to basic form. Trigonometric functions are *periodic,* which means that trigonometric equations have an infinite number of solutions unless restricted by a *side condition.* Solutions must satisfy the equation as well as the side condition.

Graphing

Trigonometry involves extensive work with the graphical representation of *sine curves, cosecant curves, cosine curves, secant curves, tangent curves,* and *cotangent curves* as well as the domains and ranges of points plotted on such curves.

Vectors

A *vector quantity* has size or magnitude as well as direction, in contrast to a *scalar quantity*, which has only size or magnitude. Velocity is an example of a vector quantity, because it incorporates magnitude and direction, whereas speed is scalar, an expression of magnitude without regard to direction.

Vector operations are some of the most useful and interesting operations in trigonometry and have a host of practical applications. Vector operations are essential in physics, particularly classical mechanics.

Polar Coordinates

For much of the introductory-level trigonometry course, you will work with the rectangular coordinate system described in Chapter 9. Another system is the *polar coor-*

dinate system, which consists of a fixed point *0*, which is the *pole* or *origin*. Extending from the pole is the *polar axis*, a ray situated horizontally and to the right of the pole. Any point *P* in the plane can be located by specifying an angle and a distance.

◆ The angle *θ* is measured from the polar axis to a line passing through the point and the pole. If measured in a counterclockwise direction, the angle is positive. If measured in a clockwise direction, it is negative.

◆ The directed distance *r* is measured from the pole to point *P*. If *P* is on the terminal side of the angle, *r* is positive. If *P* is on the opposite side of the pole, *r* is negative.

◆ As with points on the rectangular coordinate system, points on the polar coordinate system can be written as ordered pairs (r, $θ$).

Complex Numbers

Complex numbers combine real and imaginary numbers (imaginary numbers are the square roots of negative numbers) and are written in the form $a + bi$, where a and b are real numbers. Polar coordinates are often used to graph complex numbers on the *complex plane*, where the x-axis is the *real axis* and the y-axis is the *imaginary axis*.

Working with complex numbers may come late in an introductory trigonometry course and is also important preparation for calculus.

Limits

Another topic introduced toward the end of a trig course is *limits*, central to calculus, which use the limit concept to arrive at results. In trigonometry, limits are especially important in working with the areas of sine curves.

CHAPTER 11

CALCULUS OVERVIEW

T HE MOST CHALLENGING AND FASCINATING OF THE CORE AREAS OF MATHEMATICS IS calculus. Although this branch of mathematics is rooted in the work of Archimedes (287?–212 BC) and even earlier Greek mathematicians (from the schools of Plato and Eudoxos), Sir Isaac Newton (1642–1727) and Gottfried Wilhelm Leibniz (1646–1716) established the field in essentially its present form.

At the heart of calculus is the concept of the limit, which is used to arrive at a result. It was this concept that Archimedes used to find the area of the circle as πr^2. Archimedes inscribed equilateral polygons in a circle. As the number of sides of the polygon increased, the area of the polygon (readily calculable) approached the area of the circle as a limit. Combining this result with a similar idea involving *circumscribed* polygons, Archimedes arrived at the πr^2 formula. Called by later mathematicians the "method of exhaustion," the idea of confining an unknown area or volume within two known or calculable quantities, one increasing toward the unknown and the other decreasing toward it, became well established.

Extending this method of exhaustion, it is possible to calculate the area of an irregularly shaped plate by subdividing it into rectangles of equal width. As the number of rectangles is made larger, the sum of their areas approaches the unknown area as a limit. This method can be extended to solids as well. In part, calculus provides a systematic way to use limits to calculate areas, volumes, and other quantities.

Among the most valuable applications of calculus is physics, in particular calculations of acceleration. Legend has long held that Newton was moved to discover calculus by watching an apple fall from a tree. He observed that the apple had not only velocity but acceleration. To express this *continuous* quantity mathematically, Newton proposed a way of applying limits to the continuous quantity. He supposed that at any stage of its motion, the apple drops an *additional* distance Δs during a brief *additional* interval of time Δt. Velocity, therefore, nearly equals Δs divided by Δt. The *exact* velocity would be the *limit* of $\Delta s / \Delta t$ as Δt approaches 0:

$$v = \lim_{\Delta t \to 0} \frac{\Delta s}{\Delta t} = \frac{ds}{dt}$$

The quantity ds/dt is the rate of change of s with respect to t; that is, ds/dt is the *derivative* of s with respect to t. We may think of ds and dt as numbers whose ratio ds/dt is equal to v; ds is the *differential* of s, and dt is the *differential* of t.

As velocity is the derivative (rate of change) of the distance with respect to time, so acceleration is the derivative (rate of change) of the velocity with respect to time. So acceleration a may be calculated:

$$a = \frac{dv}{dt} = \lim_{\Delta t \to 0} \frac{\Delta v}{\Delta t}$$

where Δv is the increase in velocity that occurs during Δt. Since a is the derivative of v, and v is the derivative of s, a is also the *second derivative* of s:

$$a = \frac{dc}{dt} = \frac{d}{dt}\left(\frac{ds}{dt}\right) = \frac{d^2 s}{dt^2}$$

To find the derivatives of s with respect to t, we must know the dependence of s on t; that is, s must be expressed as a function of t. *Differential calculus* is that part of calculus that deals with derivatives, the instantaneous rates of change of continuous functions.

So, if we are given s as a function of t, we can find the derivative v of s. The converse is also true. If v is known, we can work back to calculate s. The basic equation $v = ds/dt$ is rewritten as $ds = vdt$. The quantity s, which we are to find, is regarded as the *antidifferential* of ds and is denoted by an integral sign in an equation that specifies s the integral of v with respect to t:

$$\int ds = \int vdt \quad \text{or} \quad s = \int vdt$$

The part of calculus that deals with integrals, finding functions when their derivatives are known, is called *integral calculus*. The British mathematician Isaac Barrow (1630–1677) was the first to recognize that differentiation and integration are inverse operations.

While calculus is an advanced form of mathematics, it is hardly limited to math majors. It is of special interest to students planning on majors in the physical sciences—especially physics and engineering—and in business (many universities offer courses specifically in business calculus).

Pre-Calculus

Most mathematics departments offer diagnostic placement tests to help determine whether calculus is appropriate for you. Talk to a math advisor or the instructor of the prospective course.

At the very least, the instructor will assume that students entering a fundamental calculus course have a solid understanding of:

◆ Interval notation, wherein (for example) $(a,b) = \{x \in R: a < x < b\}$ and $(-\infty, +\infty)$ $= \{x \in R\}$.

◆ The concept of absolute value:

$$|x| = \begin{cases} x, x > 0 \\ 0, x = 0 \\ -x, x < 0 \end{cases}$$

or

$$|x| = \sqrt{x^2}$$

◆ Functions and trigonometric functions (see Chapter 10).

◆ Algebra background, including a high comfort level with linear equations.

The Limit Concept

The concept most fundamental to calculus is the *limit*. Generations of mathematicians have created formal definitions of the limit concept, but in introductory calculus an intuitive definition suffices. It is essentially this: the limit of a function $f(x)$ describes the behavior of the function close to a particular value of x.

$$\lim_{x \to c} f(x) = L$$

As the independent variable x approaches c, the function value $f(x)$ approaches the real number L. If the function does not approach a real number L as x approaches c, the limit does not exist:

$$\lim_{x \to c} f(x) \text{DNE}$$

To determine the numeric value of limits (to evaluate limits), you will use techniques ranging from pattern recognition to substitution to algebraic simplifications.

Limits may be *one-sided*, with x approaching c from the left or right only, or *infinite*, increasing or decreasing without bound. As introduced in trigonometry (see Chapter 10), the trigonometric functions sine and cosine have important limit properties, which can be used to evaluate limit problems involving the six basic trig functions.

Functions may be *continuous* or *discontinuous*. A function is continuous in an interval if it has no gaps, splits, or holes in the given interval. Such common functions as linear, quadratic, and other polynomial functions, rational functions, and trigonometric functions are continuous at each point in their domains. Many problems in calculus apply the limit concept to the concept of continuity in order to *discuss* the continuity of a function at a given point.

The Derivative

The concept of the derivative was defined earlier as the rate of change of a function (more precisely, as the instantaneous rate of change of a function at a point). The derivative of a function at a point may also be thought of as:

◆ The slope of the tangent line at that point
◆ The instantaneous velocity of a function representing the position of a particle along a line at time t, where $y = s(t)$

Differentiation

Finding derivatives is made easier by learning a set of *differentiation rules* and applying them. In the case of the six trigonometric functions, special differentiation formulas apply and can be used in application problems of the derivative.

Composite Functions

When working with a *composite function*—that is, a function of a function, in which f is the external function and g is the core function—the *chain rule* provides a technique for finding the derivative. The exams in Chapters 20 through 22 include problems in which the chain rule applies. The chain rule is also applied in cases of *implicit differentiation*, equations in x and y that do not explicitly define y as a function of x.

Derivative Applications

The derivative of a function may be used to sketch curves, solve maximum and minimum problems, solve typical mechanics problems (distance, velocity, acceleration), solve related rate problems, and approximate function values.

Key subjects in the application of the derivative include:

◆ Finding equations of tangent and normal lines at a given point

◆ Finding *critical points*—points on the graph of a function where the derivative is 0 or does not exist

◆ Determining possible minimum and maximum values of a function on certain intervals by application of the *Extreme Value Theorem*

◆ Application of the *Mean Value Theorem*

◆ Determining whether a function is increasing or decreasing on any intervals in its domain

◆ Analysis of curves, including monotonicity and concavity

◆ Optimization of global as well as local extrema

◆ Modeling rates of change

Integral Calculus

After covering differentiation (differential calculus), the course moves on to "anti-differentiation" or *integration* (integral calculus), the inverse of differentiation. (Note that some introductory calculus courses are brief courses, while others may be given over a two- or even three-semester sequence.) If the rate of change of a function is known, integration enables us to determine the nature of the original function:

$$\int f(x)dx$$

Integral calculus involves working with *indefinite integrals*, in which the limits of integration are not specified (as in the preceding expression), and with *definite integrals*, in which the limits of integration are specified in the form:

$$\int_a^b f(x)dx$$

In addition to working with the basic properties of integrals, you will become familiar with a host of basic formulas and techniques for integration.

Application of Indefinite Integrals

Indefinite integrals can be used in a variety of applications to model physical as well as social and economic situations. Problems of distance, velocity, and acceleration, including the case of the free-falling object, are common applications of the indefinite integral. Typically, this part of the course will involve some real-world applications of calculus, with an emphasis on using the integral of a rate of change to give accumulated change.

The Definite Integral and the Reimann Sum

The concept of the definite integral can best be understood as an integral defined between two values *a* and *b* of an independent variable. Visualized as a graph, the definite integral is the area contained between the graph of a function and the *x*-axis. The *Reimann sum* is the sum of the areas of rectangles used to approximate the area between the graph and the *x*-axis.

The Fundamental Theorem of Calculus

The problem with using Reimann sums to evaluate definite integrals is that evaluating the limit of a Reimann sum can be very difficult as well as time consuming. The Fundamental Theorem of Calculus not only establishes the relationship between indefinite and definite integrals, but provides a technique for evaluating definite intervals without resorting to Reimann sums. By the conclusion of the introductory calculus course, you will be using the Fundamental Theorem to evaluate definite integrals and to represent a particular antiderivative.

Applications of Definite Integrals

The definite integral of a function is most frequently applied to problems of:

◆ Areas bounded by curves

◆ Volumes of solids by slicing (volumes of solids with known cross-sections)

◆ Volumes of solids of revolution (disks, washers, cylindrical shells)

◆ Lengths of arcs of a curve

Problems involving applications of this nature typically conclude the introductory calculus course.

PART THREE

MIDTERMS AND FINALS

ALGEBRA

CHAPTER 12

BRIGHAM YOUNG UNIVERSITY

MATH 110: COLLEGE ALGEBRA

Jonathan Andrew Bodrero, Graduate Student Instructor

THREE MIDTERMS AND ONE FINAL ARE GIVEN IN THIS COURSE.

Primary course objectives are to:

1. Help students to overcome the "math fear" and help them learn and excel.

2. Teach the basic concepts of algebra.

3. Encourage students to continue learning about mathematics by taking more advanced courses.

In addition to overcoming the math fear (it's really not all that bad!), I want my students to

1. See that math (algebra) is used in the real world and that it is worth their time to learn it.

2. Learn basic concepts of algebra: root-finding, definitions, handling story problems, etc.

3. Develop the ability to think abstractly (although this is only limited abstraction).

The exam problems have been chosen to help facilitate the goals and skills just mentioned. The story problems should serve to show the student that math is used in the real world and let him/her know that he/she generally won't come across non-story problems in real life. Learning how to decipher them is a key skill.

"Proper prior planning prevents poor performance." I don't know who first said this, but the fact is that the student who works consistently throughout the semester will perform well on the exams and learn that math isn't all that bad and that it even has practical uses.

The exam questions test the student's knowledge of definitions and the other basic concepts of algebra such as root-finding, etc. Some of the problems may require the student to think abstractly and come up with new ways of combining tools to solve problems.

MIDTERM EXAM

Time limit: 50 min. NO CALCULATORS ALLOWED.

Part I: Definitions

1. Function –
2. Range of a function –
3. Domain of a function –
4. Quadratic equation –
5. "Zero" of a polynomial –
6. Direct variation –
7. Inverse variation –
8. Joint variation –
9. Even function –
10. Odd function –

Answers:

1. A rule or mapping from one set to another that satisfies the vertical line test
2. Outputs of a function
3. Inputs of a function
4. Polynomial of degree
5. A number (possibly complex) that satisfies a given polynomial
6. $y = kx$ for some $k \neq 0$
7. $y = k/x$ for some $k \neq 0$
8. $z = kxy$ for some $k \neq 0$
9. $f(-x) = f(x)$
10. $f(-x) = -f(x)$

Part II: Written Problems

1. Find the equation (2 pt. form) of the line that goes through the points (4,3) and (–4,–4). Graph the line.

Answer:

$$y - 3 = \frac{-4-3}{-4-4}(x-4) \quad \Leftrightarrow \quad y - 3 = \frac{7}{8}(x-4)$$

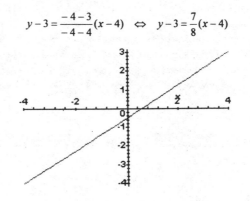

2. Find the equation (slope-intercept form) of the line that goes through the points

$$(2, \frac{1}{2}) \text{ and } (\frac{1}{2}, \frac{5}{4})$$

Graph the line.

Answer:

$$y = \frac{-1}{2}x + \frac{3}{2}$$

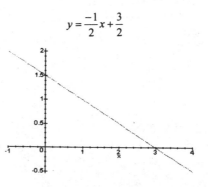

3. Is this a function? If so, what is the domain of the function? Find the value of the function at the indicated point.

a) $f(x) = \sqrt{4 - x^2}$ $\qquad\qquad$ $f(1) = ?$

Answer:

Yes [–2,2]

$$f(1) = \sqrt{3}$$

Realize that the function is only defined when the number under the radical is greater than or equal to zero. Hence, $4 - x^2 \geq 0 \implies x^2 \leq 4 \implies -2 \leq x \leq 2$

b) $g(x) = \dfrac{1}{x+5}$ $\qquad\qquad$ $g(-1) = ?$

Answer:

Yes. All reals except $x = -5$. $g(-1) = 1/4$

Realize that it is a function and that when x = −5, the denominator is zero and not defined.

4. Are the following functions even, odd, or neither?

a) $g(x) = \dfrac{1}{x+5}$ $\qquad\qquad$ $g(-1) = ?$

Answer:

Even

g(−s) = g(s)

b) $h(x) = x^3 - 5$

Answer:

Neither

x^3 is odd, while $5 - 5x^0$ is even. Hence it is neither odd nor even.

5. Graph the following functions.

a) $f(x) = |x| + 1$

Remember what the graph of f (x) = |x| looks like, and realize that adding the +1 to the absolute value function shifts it UP one unit.

Answer:

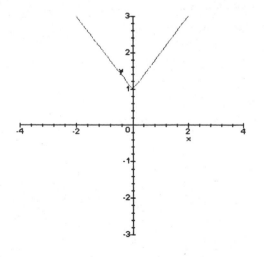

b) $g(x) = |x + 1|$

Recall the graph of the absolute value of x, and realize that the change here shifts the graph to the LEFT one unit.

Answer:

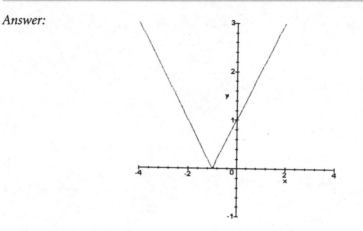

c) $h(x) = |x + 1| - 2$

Recall the graph of the absolute value of x and that the +1 inside the absolute value shifts it to the LEFT one unit. Remember that the −2 at the end shifts the entire graph DOWN two units.

Answer:

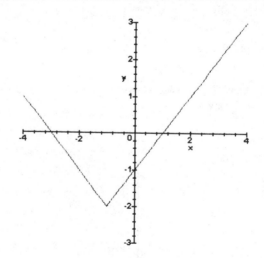

6. Show that $f(x) = 3 - 4x$ and $g(x) = \dfrac{3-x}{4}$ are inverses.

Answer:

　　1) Let $y = f(x)$.

　　2) Swap x and y.

　　3) Solve for y.

　　4) Replace y by $f^{-1}(x)$.

　　5) Make sure domain and range of $f(x)$ match the range and domain of f inverse.

7. For the following function, does $f^{-1}(x)$ exist? If so, find it. If not, explain why not.
$$f(x) = 1 - x^3$$

Answer:

　　Inverse exists because the function passes the vertical line test.

$$f^{-1}(x) = \sqrt[3]{1-x}$$

1) Let $y = f(x)$.
2) Swap x and y.
3) Solve for y.
4) Replace y by $f^{-1}(x)$.
5) Make sure domain and range of $f(x)$ match the range and domain of f inverse.

8. Factor the following polynomials and find the zeros of the polynomials.

　　a) $f(x) = x^2 + x - 2$

Answer:

　　$f(x) = (x+2)(x-1)$ Hence the zeros are $x = -2$ and $x = 1$.

　　b) $g(t) = t^5 - 6t^3 + 9t$

Answer:

$$t(t^2 - 3)(t^2 - 3) \quad \text{Hence the zeros are } t = 0, \quad t = \pm\sqrt{3}$$

Factor out the t, then factor the perfect square OR use quadratic formula.

9. Find a polynomial with the following zeros: $-2, -1, 0, 1, 2$. (Please write the polynomial in the linear combination form and the regular form.)

Answer:

　　$(x+2)(x+1)x(x-1)(x-2)$ and $x^5 - 5x^3 + 4x$

10. Divide by LONG DIVISION:

a) $\dfrac{7x+3}{x+2}$

Answer:

$7R-11$

b) $\dfrac{x^4+3x^2+1}{x^2-2x+3}$

Answer:

$x^2+2x+4 \ \ R2x-11$

Know long division!

11. Divide by SYNTHETIC DIVISION:

a) $\dfrac{5x^3-6x^2+8}{x-4}$

Answer:

$5x^2+14x+56 \ \ R232$

b) $\dfrac{5x^3}{x+3}$

Answer:

$5x^2-15x+45 \ \ R-135$

Know synthetic division, too!

MIDTERM EXAM / TAKE-HOME PORTION

No Time limit

NO CALCULATORS ALLOWED

1. Find the following for $f(x)=\dfrac{x}{x+1}$ and $g(x)=x^3$

a) $(f+g)(x)$

Answer:

$$\dfrac{x^4+x^3+x}{x+1}$$

b) $(g+f)(x)$

Answer:

$$\dfrac{x^4+x^3+x}{x+1}$$

c) $(f-g)(x)$

Answer:

$$\dfrac{7x+3}{x+2}$$

d) $(g-f)(x)$

Answer:

$$\frac{x^4 + x^3 - x}{x+1}$$

e) $(fg)(x)$

Answer:

$$\frac{x^4}{x+1}$$

f) $(gf)(x)$

Answer:

$$\frac{x^4}{x+1}$$

g) $(f/g)(x)$

Answer:

$$\frac{x}{x^4 + x^3}$$

1) Vertex of a parabola is $\dfrac{-h}{2a}$ so vertex when $x = \dfrac{2}{2(-\frac{1}{12})} = \dfrac{-2}{\frac{-1}{6}} = (-2)(-6) = 12$

2) Plugging in $x = 12$ yields $y = 16$. Vertex can also be found by placing the equation in standard form (requires completing the square).

h) $(g/f)(x)$

Answer:

$$x^3 + x$$

i) $f \circ g(x)$

Answer:

$$\frac{x^3}{x^3 + 1}$$

j) $g \circ f(x)$

Answer:

$$\frac{x^3}{(x+1)^3}$$

Know how to add, subtract, multiply, divide, and compose polynomials.

2. The height y in feet of a ball thrown by a child is given by $y = \frac{-1}{12}x^2 + 2x + 4$ where x is the horizontal distance (in feet) from where the ball is thrown.

a) How high is the ball when it leaves the child's hand? (Note: Find y when $x = 0$.)

Answer:

4 ft.

Use the hint given in the problem.

b) How high is the ball when it is at its maximum height?

Answer:

16 ft.

1) Vertex of a parabola is $\frac{-b}{2a}$ so vertex when $x = \frac{-2}{2(-\frac{1}{12})} = \frac{-2}{\frac{-1}{6}} = (-2)(-6) = 12$

2) Plugging in $x = 12$ yields $y = 16$. Vertex can also be found by placing the equation in standard form (requires completing the square).

c) How far from the child does the ball strike the ground?

Answer:

$$x = 12 + 8\sqrt{3}$$

When the ball hits the ground, $y = 0$. Use the quadratic equation to find that $y = 0$. The student should realize that this expression—

$$x = 12 - 8\sqrt{3}$$

is also a solution to the quadratic equation, but it is a negative number and not the right answer.

3. Do the following for $f(x) = -4x^3 + 15x^2 - 8x - 3$:

a) List all the possible rational zeros for the following function.

Answer:

$$\frac{\pm 3 \quad \pm 1}{\pm 1 \quad \pm 2 \quad \pm 4}$$

Know the rational zero test. There's no other way around it.

b) Sketch a graph of the function

Either create a table using several points such as $x = -1, 0, 1, 2, 3$ and sketch (crude but effective), or find the zeros (which you will use to answer part c), and then plot intermediate points to be able to sketch the rest of the graph.

Answer:

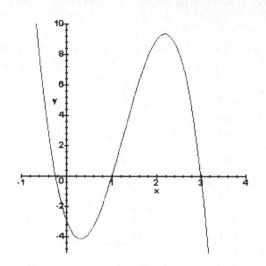

c) Determine all real zeros.

Answer:

$$\frac{-1}{4}, 1, 3$$

Inspection (or the graph) reveals that $x = 1$ is a zero. Factor this out. To find the other zeros, use the quadratic formula on the remaining polynomial.

4. An open box is to be made from a rectangular piece of material, 15 centimeters by 9 centimeters, by cutting equal squares from the corners and turning up the sides.

a) Let x represent the length of the sides of the squares removed. Draw a figure showing the squares removed from the original piece of material and the resulting dimensions of the open box.

Answer:

The figure should look like a box with two vertical lines and two horizontal lines drawn parallel to the sides of the box, but inside the box, with each line being a distance x from the corresponding side.

b) Use the figure to write the volume V of the box as a function of x. Determine the domain of the function.

Answer:

$V =$ Length $\times$ Width $\times$ Height. Hence, $V = (15 - 2x)(9 - 2x)x$.

Since this is just a cubic polynomial, the domain of the function is the entire real line.

c) Sketch the graph of the function and approximate the dimensions of the box that yield a maximum volume.

Answer:

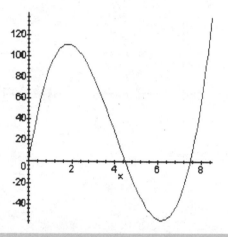

Create a table to get a rough idea of the graph. Values for x could include: $x = 0, 1, 2, 3, 4, 6, 8$. Also, the student should realize the right-hand and left-hand behavior of the function, since it is just a cubic polynomial (goes to $\pm\infty$, respectively). From the graph, we see that the function is maximized around $x = 2$ with a value of $y = 110$.

d) Find values of x such that $V = 56$. Which of these values is a physical impossibility in the construction of the box? Explain.

Answer:

Solving the polynomial yields $\frac{1}{2}, \frac{7}{2}, 8$

Use the rational zeros to pick one root to divide out of the polynomial.

If you divide the polynomial by the root found in part 2, the polynomial reduces to degree two and can be solved by using the quadratic formula. For example: if you divide by $x =$ and then use the quadratic formula, the other roots are easily found.

$x = 8$ is the impossible one because $9 - 2x = -7$, and you can't have a box with a width of -7.

5. Find a polynomial with INTEGER coefficients that has the given zeros. Write out the polynomial in linear factor form and regular form.

a) $6, -5 + 2I, -5 - 2i$

Answer:

$$(x^2 + 10x + 29)(x - 6) \text{ and } x^3 + 4x^2 - 31x - 174$$

Multiply the two complex conjugates together to get the quadratic factor seen above, then include the linear factor and multiply the polynomial out to get it in standard form.

b)

$$\frac{3}{4}, -2, \frac{-1}{2} + i$$

Answer:

$$16x^4 + 36x^3 + 16x^2 + x - 30$$

$$16\left(x^2 + x + \frac{5}{4}\right)\left(x - \frac{3}{4}\right)(x + 2)$$

1) Multiply the complex number by its conjugate (because they occur in pairs for polynomials with real—integer—coefficients). This yields the quadratic factor seen in the solution.
   This is a key point that is being tested here. Part a should remind the student of this point if it has slipped his/her mind.

2) Multiply that quadratic polynomial by $\left(x - \frac{3}{4}\right)$ and (x + 2) for the other roots.

3) The problem requires that the coefficients must be integers, so multiply by 16 (which will not change the roots, only stretch the graph vertically).

Time Limit: 2 hours 50 min.

NO GRAPHING CALCULATORS ALLOWED

Advice:
1. READ THE BOOK!
2. Come to class PREPARED to ask questions and take notes.
3. UNDERSTAND and complete each homework assignment.
4. Review for the exams.
5. Fulfill other course work as outlined in my syllabus.
6. Have fun!

This exam has a multiple choice and a written section. On the multiple choice section, please write your answer in the blank next to the question number. On the written section of the exam, please show ALL of your work on the test.

Multiple Choice Section

1. The graph below is a transformation of the graph of $f(x) = |x|$. Find the equation which most closely resembles the graph.

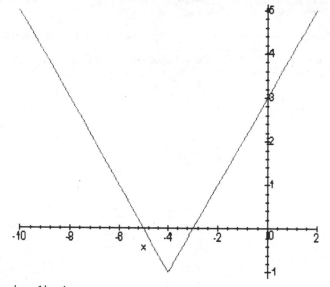

    (a) $g(x) = |x - 1| - 4$

    (b) $g(x) = |x - 4| - 1$

    (c) $g(x) = |x - 1| + 4$

    (d) $g(x) = |x + 4| - 1$

Answer: b

1) Graph $f(x) = |x|$.
2) Realize what kind of translations are necessary to move it left 4 and down 1 ($|x - 4|$ and -1, respectively).
3) Apply the corresponding translations.

2. An open box is made from a rectangular piece of material by cutting equal squares from each corner and turning up the sides. Write the volume of the box as a function of x if the material is 24 inches by 16 inches.
   (a) $V = (24 - x)(16 - x)$
   (b) $V = x(24 - x)(16 - x)$
   (c) $V = x(24 - 2x)(16 - 2x)$
   (d) $V = (24 - 2x)(16 - 2x)$
   (e) None of the above.

Answer: c

1) Draw a graph of the box, labeling everything.
2) Imagine the box folded up.
3) Recall the formula for calculating the volume of a box
   (Volume = Length × Width × Height; V = LWH).
4) Use the corresponding formula with the correct measurements. This can be trickier than you think, because you have to take x from each side to fold up the box; hence the length is $24 - 2x$ and the width is $16 - 2x$.

3. Classify the graph of $2x^2 - 5y^2 + 4x - 6 = 0$.
   (a) Circle
   (b) Hyperbola
   (c) Ellipse
   (d) Parabola
   (e) None of these.

Answer: b

1) Complete the square for this conic, and write it in standard form.

$$2x^2 - 5y^2 + 4x - 6 = 0 \implies 2(x^2 + 2x + 1) - 2 - 5y^2 - 6 = 0 \implies$$

$$2(x+1)^2 - 5y^2 - 8 = 0 \implies \frac{(x+1)^2}{5} - \frac{y^2}{2} - 8 = 0$$

2) Recognize the standard form resembles that of a hyperbola.
Tip: If you blank on the formula or can't complete the square, plotting 4 to 6 points can help decipher the shape of the conic. Then you need only to remember the name of it.

4. Find the domain of the function $h(x) = \sqrt{\dfrac{x+4}{x(x-5)}}$

    (a) $(-\infty,0), (0,5), (5,\infty)$

    (b) $(-\infty,-4), (-4,0), (0,5), (5,\infty)$

    (c) $(-4,0), (0,5), (5,\infty)$

    (d) $(-\infty,-4), (-4,5), (5,\infty)$

    (e) None of the above.

Answer: e

1) Realize that there are vertical asymptotes at $x = 0$ and $x = 5$ (where the denominator is equal to zero).

2) Realize that the function is not defined for negatives under the radical, hence any number less than −4 is not in the domain of the function. This eliminates 3 of the choices. Inspection shows that C cannot be correct because $h(1)$ is not in the domain either. Hence the correct answer is e, None of the above.

   If you forget about asymptotes, plugging in points from each piece of each choice will show that none of the given answers are correct and that e must be the correct choice. This method is slow but still effective.

5. Find an equation of the line that passes through (6,2) and is perpendicular to the line $3x + 2y = 2$

    (a) $y = \dfrac{-3}{2}x + 11$

    (b) $y = \dfrac{-2}{3}x + 6$

    (c) $y = \dfrac{3}{2}x - 7$

    (d) $y = \dfrac{2}{3}x - 2$

    (e) None of the above.

Answer: d

1) Know that perpendicular means that $m_1 = \dfrac{-1}{m_2}$

2) Use the point-slope formula: $y - y_1 = m_1(x - x_1)$ $\Rightarrow$ $y - 2 = \dfrac{2}{3}(x - 6)$ and the answer follows.

6. Is the following function even or odd? $f(x) = 4x^3 + 3x$
   (a) Even
   (b) Odd
   (c) Both
   (d) Neither

Answer: b

> Noticing that the exponents are all odd, 3 and 1, you should realize that the function is odd.

7. $F(x) = \begin{cases} 3 - x^2, & if \ x \geq 0 \\ 3 + 2x, & if \ x < 0 \end{cases}$ $F(2) - F(1) = ?$

   (a) −6
   (b) 0
   (c) −2
   (d) $F(3)$
   (e) None of the above.

Answer: e

> 1) Realize that the function is split and only evaluate the points on the part of the function where it is defined.
> 2) $F(2) - F(1) = -1 - 2 = -3$
> 3) Since −3 is not one of the choices ($F(3) = -6$), the only choice left is e, None of the above.

8. T varies directly with the square root of L and inversely with the square root of g. If $T = \frac{\pi}{2}$ when $L = 2$ and $g = 32$, find the constant of proportionality.

   (a) 2π
   (b) $\pi/8$
   (c) π
   (d) $\pi/2$
   (e) None of the above.

Answer: a

> 1) Write the words out into a formula: $T = k\dfrac{\sqrt{L}}{\sqrt{g}}$, where k is the constant of proportionality.
>
> 2) Plug in the known values and solve for k as illustrated below and the answer follows:
>
> $$\frac{\pi}{2} = \frac{k\sqrt{2}}{\sqrt{32}} \Rightarrow k = \frac{\sqrt{32}\ \pi}{2\sqrt{2}} \Rightarrow k = \frac{\sqrt{16}\ \pi}{2} \Rightarrow k = \frac{4\pi}{2}$$
>
> 3) Note that this problem requires the student to remember the rules of square roots.

9. Find the x- and y-intercepts: $y = x^2 - 5x + 4$
    (a) $(0,-4), (0,1), (4,0)$
    (b) $(0,4), (4,0), (1,0)$
    (c) $(0,-4), (-4,0), (-1,0)$
    (d) $(0,4), (-4,0), (-1,0)$
    (e) None of the above.

Answer: b

1) Set y equal to zero and then use the quadratic equation to find x-intercepts:

$$\frac{5 \pm \sqrt{(-5)^2 - 4(1)(4)}}{2(1)} \Rightarrow \frac{5 \pm \sqrt{25 - 16}}{2} \Rightarrow \frac{5 \pm 3}{2} \Rightarrow 1, 4$$

2) Sub $x = 0$ into the equation to find out the y-intercept:
$$y = 0 - 0 + 4 \Rightarrow y = 4$$

10. Write in the form $y = a(x - h)^2 + k$: $y = -2x^2 - 4x - 5$.
    (a) $y = -2(x - 1)^2 - 2$
    (b) $y = (2x - 2)^2 - 1$
    (c) $y = -2(x + 2)^2 - 1$
    (d) $y = -2(x + 1)^2 - 3$
    (e) None of the above.

Answer: d

Complete the square, and write in standard form.

11. The graph most closely resembles which function?

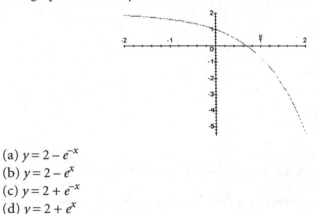

    (a) $y = 2 - e^{-x}$
    (b) $y = 2 - e^{x}$
    (c) $y = 2 + e^{-x}$
    (d) $y = 2 + e^{x}$

Answer: b

1) Know what the graphs of e^x and e^{-x} look like.
2) Realize that the graph is most similar to $-e^x$.
3) Realize that there appears to be a horizontal asymptote at $y = 2$.
4) Make the appropriate shift.

12. Given $f(x) = \dfrac{1}{x^2}$ and $g(x) = \sqrt{x^2 + 4}$, find $(f \circ g)(x)$

(a) $\dfrac{1}{x^2 + 4}$

(b) $\dfrac{1}{\sqrt{x^2 + 4}}$

(c) $x^2 + 4$

(d) $\dfrac{1}{x\sqrt{x^2 + 4}}$

(e) None of the above.

Answer: a

Know composition of functions, plug them in, and go.

13. Find the vertical asymptote(s) for $f(x) = \dfrac{x + 3}{(x - 2)(x + 5)}$
   (a) $y = 2, y = -5, y = -3$
   (b) $x = 2, x = -5, x = -3, x = 1$
   (c) $x = 1$
   (d) $x = 2, x = -5$
   (e) None of the above.

Answer: d

Know that vertical asymptotes are when the denominator goes to zero, hence $x = 2$, $x = -5$.

14. Determine the amount of money that should be invested at a rate of 6.5% compounded monthly to produce a final balance of $15,000 in 20 years.
    (a) $4102.34
    (b) $5216.07
    (c) $2458.83
    (d) $14,056.14
    (e) None of the above.

Answer: a

> 1) Know the interest formula: $A = P\left(1 + \dfrac{r}{n}\right)^{nt}$
>
> where A is the amount after interest is calculated, r is the rate of interest changed to a decimal form, n is the number of times it is compounded each year and t is the number of years.
>
> 2) Solve for P: $P = \dfrac{A}{\left(1 + \dfrac{r}{n}\right)^{nt}}$
>
> 3) Insert the values and calculate P.

15. Write the exponential form of $\log^b 7 = 13$
    (a) $7^{13} = b$
    (b) $b^{13} = 7$
    (c) $b^7 = 13$
    (d) $7^b = 13$
    (e) None of the above.

Answer: b

> Know the definition of logs: $\log^x y = z \Leftrightarrow x^z = y$.

16. Given $A = \begin{bmatrix} 3 & 6 & -1 \\ 0 & 5 & 2 \end{bmatrix}$ and $B = \begin{bmatrix} 1 & 0 & 5 \\ -1 & 2 & 7 \end{bmatrix}$, find $3A - 2B$

    (a) $\begin{bmatrix} 7 & 18 & -13 \\ 2 & 11 & -8 \end{bmatrix}$

    (b) $\begin{bmatrix} 7 & 18 & 2 \\ 0 & 11 & -8 \end{bmatrix}$

    (c) $\begin{bmatrix} 11 & 18 & 7 \\ -2 & 19 & 20 \end{bmatrix}$

    (d) $\begin{bmatrix} 7 & 18 & -13 \\ -2 & 9 & 20 \end{bmatrix}$

    (e) None of the above.

Answer: a

Know scalar multiplication of a matrix; know how to add (subtract) matrices.

17. Find the 99th term of the arithmetic sequence with $a^1 = 7$ and $d = -3$. (Assume that n begins with 1.)
    (a) –287
    (b) –290
    (c) –293
    (d) –297
    (e) None of the above.

Answer: a

Know the formula for the *n*th term of an arithmetic sequence:
$a^n = a^1 + (n - 1)d.$

18. There are seven possible digits in a phone number. How many different phone numbers are possible if the first digit cannot be 0 and no digit can be used more than once?
    (a) 128
    (b) 181,440
    (c) 544,320
    (d) 5,040
    (e) None of the above.

Answer: c

Consider each digit left to right. Recognize when the digits have already been used. The first digit cannot be 0, so it has 9 different possibilities. The second cannot be the same as the first, which leaves 9 possibilities. The third cannot be the first or second, which leaves 8 possibilities. And so on. Then multiply it out:
$$9 \times 9 \times 8 \times 7 \times 6 \times 5 \times 4 = 544{,}320.$$

19. Find the sum of the infinite geometric sequence:

$$1, \frac{1}{3}, \frac{1}{9}, \frac{1}{27}, \ldots$$

    (a) $\dfrac{3}{2}$
    (b) 3
    (c) $\dfrac{5}{3}$
    (d) $\dfrac{5}{2}$
    (e) None of the above.

Answer: a

1) Know the formula for the summation of an infinite geometric series: $\dfrac{a_1}{1-r}$ *for* $|r|<1$.

2) Realize that $r = \dfrac{1}{3}$ in this case and the first term in the series is equal to 1.

3) Plug in, and the answer follows.

20. Find the number of distinguishable ways the letters OKEECHOBEE can be arranged.
   (a) 75,600
   (b) 3,628,800
   (c) 151,320
   (d) 1,814,400
   (e) None of the above.

Answer: a

Know how to count permutations in general and how to divide those already used out.

Written Section

Neatly show all of your work in the space provided.

1. Solve the following system using:

$$x + y = 3$$
$$x + 2y = -1$$

   (a) Gauss-Jordan Elimination

Answer:

$$x = 7 \quad y = -4$$

1) Know what Gauss-Jordan Elimination is.
2) Apply basic row operations to the augmented matrix to get it to row reduced echelon form (rref).

$$\begin{bmatrix} 1 & 1 & \vdots & 3 \\ 1 & 2 & \vdots & -1 \end{bmatrix} \Leftrightarrow \begin{bmatrix} 1 & 1 & \vdots & 3 \\ 0 & 1 & \vdots & -4 \end{bmatrix} \Leftrightarrow \begin{bmatrix} 1 & 0 & \vdots & 7 \\ 0 & 1 & \vdots & -4 \end{bmatrix}$$

3) Make sure that the leading entry in each row is 1, even though, in this case, it is by default.

(b) Inverse Matrix Method
Answer: Same answer as (a)

1) Know how to calculate the inverse of a matrix in any of several ways.

$$A = \begin{bmatrix} 1 & 1 \\ 1 & 2 \end{bmatrix} \Rightarrow A^{-1} = \begin{bmatrix} 2 & -1 \\ -1 & 1 \end{bmatrix}$$

2)
$$Ax = b \Rightarrow x = A^{-1}b \quad b = \begin{bmatrix} 3 \\ -1 \end{bmatrix}. \text{ Hence } x = \begin{bmatrix} 2 & -1 \\ -1 & 1 \end{bmatrix}\begin{bmatrix} 3 \\ -1 \end{bmatrix} \Leftrightarrow x = \begin{bmatrix} 7 \\ -4 \end{bmatrix}$$

3) Complete the matrix multiplication to get the answer.

(c) Cramer's Rule
Answer: Same answer as (a)

1) Know what Cramer's Rule is.
2) Know how to calculate the determinant of a matrix.
3) Substitute the values into the proper columns and calculate the determinants and answers.

$$\frac{\begin{vmatrix} 3 & 1 \\ -1 & 2 \end{vmatrix}}{\begin{vmatrix} 1 & 1 \\ 1 & 2 \end{vmatrix}} = \frac{6 - (-1)}{2 - 1} = \frac{7}{1} = 7 = x \quad \text{and} \quad y = \frac{\begin{vmatrix} 1 & 3 \\ 1 & -1 \end{vmatrix}}{\begin{vmatrix} 1 & 1 \\ 1 & 2 \end{vmatrix}} = \frac{-1 - 3}{2 - 1} = \frac{-4}{1} = -4$$

2. Find the inverse of $f(x) = \dfrac{2}{\sqrt{x-3}}$ Be sure to make the domain of $f^{-1}(x)$ agree with the range of $f(x)$, etc.

Answer:

$$f^{-1}(x) = \frac{4}{x^2} + 3$$

1) Let $y = f(x)$.
2) Swap x and y.
3) Solve for y.
4) Replace y by $f^{-1}(x)$.
5) Check to see that range of $f(x)$, (3, infinity), is the same as the domain of the inverse.

3. Solve for x: $1n(7 - x) + 1n(3x + 5) = 1n(24x)$

Answer:

$$-5, \quad \frac{7}{3}$$

1) Use the rules of logarithms to move all the terms on one side of the equation.
2) Exponentiate both sides of the equation.
3) Get rid of the denominator by multiplying by the correct factor.
4) Solve the quadratic equation by factoring or using the quadratic equation.

4. Use Descartes' rule of signs, rational zero test, and the theorem on bounds as aids in finding all zeros for the polynomial: $f(x) = x^4 - 5x^3 + 8x^2 - 20x + 16$

Answer:

$$1, 4, 2i, -2i$$

1) Use all three tests to find that there are no negative zeros and that $x = 1$ and $x = 4$ are zeros.
2) Then divide those factors out of the polynomial by synthetic or long division.
3) Solve the quadratic equation using the quadratic formula.

5. Graph $f(x) = x^3 + 2x^2 - 8x$

1) Realize that $x = 0$ is a root (by inspection) and factor it out.
2) Find other zeros by factoring or using the quadratic formula. Either way works, but I find factoring the remaining polynomial easier.
3) Plot a few other intermediate points.
4) Sketch the rest of the graph.

Answer:

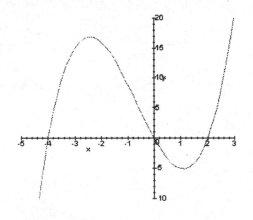

6. Graph a rational function with the following properties:
Two vertical asymptotes: $x = -3$ and $x = 2$
Horizontal asymptote: $y = 0$
One x-intercept: $(1,0)$
y-intercept: $(0,2)$

$$f(-2) = 9$$
$$f(1.5) = -2\frac{2}{3}$$
$$f(-4) = -10$$
$$f(3) = 4$$

1) Draw vertical asymptotes at $x = -3$ and $x = 2$.
2) Mark the points (1,0) and (0,2) on the graph.
3) Sketch the form of the center section using the asymptotes (the two points used in part 2), and that
$$f(-2) = 9 \text{ and } f(1.5) = -2\frac{2}{3}$$
4) Realize that the graph has a horizontal asymptote at $y = 0$ and use the fact that $f(-4) = -10$ to sketch the portion to the left of the first vertical asymptote.
5) Realize that the graph has a horizontal asymptote at $y = 0$ and use the fact that $f(3) = 4$ to sketch the portion to the right of the second vertical asymptote.

Answer:

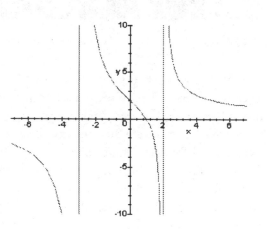

CHAPTER 13

BRIGHAM YOUNG UNIVERSITY

MATH 110: COLLEGE ALGEBRA

Nephi Noble, Graduate Student Instructor

THREE MIDTERMS AND ONE FINAL ARE GIVEN IN THIS COURSE.
Course objectives include: first, to teach a set of mathematical skills that can be applied in later math courses; second, to develop the students' ability to problem solve; third, to motivate the students as far as math goes and for learning in general; and, fourth, to build character.

I hope that students will leave this course with confidence in their ability to solve problems, math-related or not. I want my students to learn the important facets of algebra, which are required for any math courses the students may take beyond this one. Finally, students should be able to defend their thinking and be able to express this defense.

I advise students to spend time on the course. Read the text. Try to understand the concepts rather than rest content with being able only to apply an algorithm to a specific situation. Pay attention in class and actively try to anticipate what your teacher is going to do next.

MIDTERM EXAM

Part I – Rational Functions

Graph the following rational functions

1. $y = \dfrac{3x-3}{x^2+2x-3}$ Vertical Asymptotes: $x = -3$, $x = 1$ Horizontal Asymptote: $y = 0$

Answer:

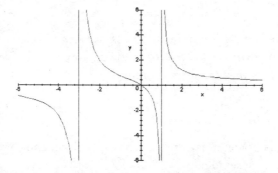

> Remember, the *x*-intercepts are found by finding the zeros of the numerator and the vertical asymptotes are found by finding the zeros of the denominator. Additionally, only one test point needs to be graphed for each section of the graph between *x*-intercepts, vertical asymptotes, and plus or minus infinity.
>
>   Note that a horizontal asymptote exists only if the degree of the denominator is less than or equal to the degree of the denominator: if less than, *y* = 0 is the horizontal asymptote. If equal to, *y* = *a/b* is the horizontal asymptote, where *a* is the leading coefficient of the numerator and *b* is the leading coefficient of the denominator.

2. $y = \dfrac{5(1-x^2)}{x^2-4}$ Vertical Asymptotes: $x = -2$, $x = 2$ Horizontal Asymptote: $y = -5$

Answer:

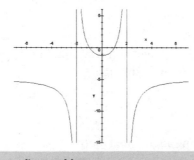

> **See comments on the preceding problem.**

Part II—Definitions

> In this part it is important to note that the memorization of definitions gives the students a base from which to build as they solve algebra problems. It would also be advantageous if the students understood these definitions as well as had them memorized.

Vertical Asymptote—
Answer:

The line $x = a$ is a vertical asymptote if, as x gets closer and closer to a, $f(x)$ gets closer and closer to ∞ or $-\infty$. (The zeros of the denominator are vertical asymptotes, and that is how they are normally found.)

Horizontal Asymptote—
Answer:

The line $y = b$ is a horizontal asymptote if, as x gets closer and closer to ∞ or $-\infty$, $f(x)$ gets closer and closer to b.

Coefficient Matrix—
Answer:

A matrix derived from the coefficients of a system of linear equations (each written in standard form with the constant term on the right) that does not include the constant terms.

Example:
$$3x + 4y = 7$$
$$-2x + y = 1$$

would give us the following coefficient matrix:

$$\begin{pmatrix} 3 & 4 \\ -2 & 1 \end{pmatrix}$$

Augmented Matrix—
Answer:

A matrix derived from a system of linear equations (each written in standard form with the constant term on the right).

Example:
$$3x + 4y = 7$$
$$-2x + y = 1$$

would yield the following augmented matrix:

$$\left(\begin{array}{cc|c} 3 & 4 & 7 \\ -2 & 1 & 1 \end{array}\right)$$

Consistent system of equations—

Answer:

A system of equations that has one or more solutions.

Inconsistent system of equations—

Answer:

A system of equations that does not have a solution.

Row Echelon Form—

Answer:

All rows that consist entirely of zeros are found at the bottom of the matrix. When a row has nonzero entries, the leftmost nonzero entry is a 1. When you have two rows that have nonzero entries, the leftmost 1 in the top of the two rows will be farther to the left than the leftmost 1 in the bottom of the two rows.

Reduced Row Echelon Form—

Answer:

It is in Row Echelon Form with the following stipulation: every column that contains a leftmost 1 of a nonzero row has zeros (the column does) in every position above and below the 1.

Gaussian Elimination—

Answer:

A process that utilizes the three basic row operations to obtain an equivalent system of equations or matrix that are in Row Echelon Form.

Gauss-Jordan Elimination—

Answer:

A process that utilizes the three basic row operations to obtain an equivalent system of equations or matrix that are in Reduced Row Echelon Form.

Identity Matrix—

Answer:

An n-by-n matrix, I, which consists of entries that are zero if they are not on the main diagonal and 1 if they are. This matrix has the unique property that if you are given an n-by-n matrix A, $AI = IA = A$.

Inverse of an n-by-n Matrix A—

Answer:

An n-by-n matrix, B, such that $AB = BA = I$. A matrix does not necessarily have an inverse. If its determinant is zero, it has no inverse. If its determinant is nonzero, it has such an inverse.

Part III—Matrices and Systems of Equations

Perform the indicated operations.

1. $\left[3\begin{pmatrix} 1 & -3 \\ 2 & 0 \end{pmatrix} + \begin{pmatrix} 3 & 3 \\ -2 & 5 \end{pmatrix}\right]\begin{pmatrix} 5 & 6 \\ -2 & 5 \end{pmatrix} = \left[\begin{pmatrix} 3 & -9 \\ 6 & 0 \end{pmatrix} + \begin{pmatrix} 3 & 3 \\ -2 & 5 \end{pmatrix}\right]\begin{pmatrix} 5 & 6 \\ -2 & 5 \end{pmatrix} = \begin{pmatrix} 6 & -6 \\ 4 & 5 \end{pmatrix}\begin{pmatrix} 5 & 6 \\ -2 & 5 \end{pmatrix} = \qquad =$

$\begin{pmatrix} (6 & -6)\begin{pmatrix} 5 \\ -2 \end{pmatrix} & (6 & -6)\begin{pmatrix} 6 \\ 5 \end{pmatrix} \\ (4 & 5)\begin{pmatrix} 5 \\ -2 \end{pmatrix} & (4 & 5)\begin{pmatrix} 6 \\ 5 \end{pmatrix} \end{pmatrix} = \begin{pmatrix} 6*5+(-6)*(-2) & 6*6+(-6)*5 \\ 4*5+5*(-2) & 4*6+5*5 \end{pmatrix} = \begin{pmatrix} 42 & 6 \\ 10 & 49 \end{pmatrix}$

Answer:

$$x + 2y + 3z = 5$$
$$2x + 3y + 5z = 7$$
$$3x + 5y + 7z = 11$$

> It is important to recognize that there is an implied matrix multiplication after the brackets and before the last 2-by-2 matrix. It is also important to remember order of operations (as well as how to matrix multiply).

2. Solve the following system of equations.

a)
$$x + 2y + 3z = 5$$
$$2x + 3y + 5z = 7$$
$$3x + 5y + 7z = 11$$

Answer:

$\begin{pmatrix} 1 & 2 & 3 & 5 \\ 2 & 3 & 5 & 7 \\ 3 & 5 & 7 & 11 \end{pmatrix} \Leftrightarrow \begin{pmatrix} 1 & 2 & 3 & 5 \\ 0 & -1 & -1 & -3 \\ 3 & 5 & 7 & 11 \end{pmatrix} \Leftrightarrow \begin{pmatrix} 1 & 2 & 3 & 5 \\ 0 & -1 & -1 & -3 \\ 0 & -1 & -2 & -4 \end{pmatrix} \Leftrightarrow \begin{pmatrix} 1 & 2 & 3 & 5 \\ 0 & 1 & 1 & 3 \\ 0 & -1 & -2 & -4 \end{pmatrix} \Leftrightarrow$

$\begin{pmatrix} 1 & 2 & 3 & 5 \\ 0 & 1 & 1 & 3 \\ 0 & 0 & -1 & -1 \end{pmatrix} \Leftrightarrow \begin{pmatrix} 1 & 2 & 3 & 5 \\ 0 & 1 & 1 & 3 \\ 0 & 0 & 1 & 1 \end{pmatrix} \Leftrightarrow \begin{pmatrix} 1 & 2 & 3 & 5 \\ 0 & 1 & 0 & 2 \\ 0 & 0 & 1 & 1 \end{pmatrix} \Leftrightarrow \begin{pmatrix} 1 & 2 & 0 & 2 \\ 0 & 1 & 0 & 2 \\ 0 & 0 & 1 & 1 \end{pmatrix} \Leftrightarrow \begin{pmatrix} 1 & 0 & 0 & -2 \\ 0 & 1 & 0 & 2 \\ 0 & 0 & 1 & 1 \end{pmatrix}$

> This can be solved without putting the system into matrix form. It turns out to be the exact same process. It is just more convenient and orderly to put it into a matrix form. This can also be solved using Cramer's rule or by using the inverse of the coefficient matrix.

b)
$$x^2 + xy = 5$$
$$y - x = 3$$

Answer:

$$y = x + 3 \rightarrow x^2 + x(x+3) = 5 \rightarrow x^2 + x^2 + 3x - 5 = 0 \rightarrow 2x^2 + 3x - 5 = 0 \rightarrow (2x+5)(x-1) = 0$$

$$x = 1 \text{ and } x = -\frac{5}{2}$$

So, $y = 1 + 3 = 4$ and $y = -\frac{5}{2} + 3 = \frac{1}{2}$

We then have two solutions : $(1,4)$ and $\left(-\frac{5}{2}, \frac{1}{2}\right)$

Part IV—Take-Home Portion—Matrices
(No Time Limit)

1. Given
$$A = \begin{pmatrix} 1 & 2 & 3 \\ 2 & 3 & 5 \\ 3 & 5 & 7 \end{pmatrix} \quad B = \begin{pmatrix} -4 & 1 & 1 \\ 1 & -2 & 1 \\ 1 & 1 & -1 \end{pmatrix}$$

Find $\det(A)$, $\det(B)$, $\det(AB)$, A^{-1}, B^{-1}.

Answer:

I'll use Dodgson's Identity for the determinant of A and the adjoint method of finding the inverse

$$\left(A^{-1} = \frac{\text{adj}(A)}{\det A} \right)$$

$$\begin{vmatrix} 1 & 2 & 3 \\ 2 & 3 & 5 \\ 3 & 5 & 7 \end{vmatrix} = \frac{\begin{vmatrix} \begin{vmatrix} 1 & 2 \\ 2 & 3 \end{vmatrix} & \begin{vmatrix} 2 & 3 \\ 3 & 5 \end{vmatrix} \\ \begin{vmatrix} 2 & 3 \\ 3 & 5 \end{vmatrix} & \begin{vmatrix} 3 & 5 \\ 5 & 7 \end{vmatrix} \end{vmatrix}}{3} = \frac{\begin{vmatrix} -1 & 1 \\ 1 & -4 \end{vmatrix}}{3} = \frac{3}{3} = 1 \text{ So } \det A = 1 \left(\text{So, } A^{-1} = \text{adj}(A) \right)$$

$$A^{-1} = \begin{pmatrix} \begin{vmatrix} 3 & 5 \\ 5 & 7 \end{vmatrix} & -\begin{vmatrix} 2 & 3 \\ 5 & 7 \end{vmatrix} & \begin{vmatrix} 2 & 3 \\ 3 & 5 \end{vmatrix} \\ -\begin{vmatrix} 2 & 3 \\ 5 & 7 \end{vmatrix} & \begin{vmatrix} 1 & 3 \\ 3 & 7 \end{vmatrix} & -\begin{vmatrix} 1 & 3 \\ 2 & 5 \end{vmatrix} \\ \begin{vmatrix} 2 & 3 \\ 3 & 5 \end{vmatrix} & -\begin{vmatrix} 1 & 3 \\ 2 & 5 \end{vmatrix} & \begin{vmatrix} 1 & 2 \\ 2 & 3 \end{vmatrix} \end{pmatrix} = \begin{pmatrix} -4 & 1 & 1 \\ 1 & -2 & 1 \\ 1 & 1 & -1 \end{pmatrix} = B \text{ (So, } AB = 1)$$

$$\det AB = \det I = 1$$
$$1 = \det AB = \det A \det B = \det B$$
As above, $B^{-1} = A$

There are various ways of finding the determinant of a matrix 3-by-3 or bigger. It is a matter of preference for some. I prefer to use Dodgson's Identity. This can only be used if the determinant of the matrix with its outside rows and columns of entries peeled off is nonzero. If it is zero, another method must be employed.

Part of the usefulness of this problem is that after finding the inverse of A or B, one should readily recognize that A and B are inverses of one another. This greatly simplifies the problem! You need only find the inverse and determinant of A and all else is simple calculation.

2. Find the determinant of the following matrices:

a)
$$\begin{pmatrix} 34z & 92y & 3757 \\ 0 & 0 & 5x \\ 0 & 0 & -12 \end{pmatrix}$$

Answer:

$$\begin{vmatrix} 34z & 92y & 3757 \\ 0 & 0 & 5x \\ 0 & 0 & -12 \end{vmatrix} = 34z * 0 * (-12) = 0$$

This problem is to see if the student recognizes that the matrix is in upper triangular form. The determinant of such a matrix is simply the product of the diagonal entries.

b)
$$\begin{pmatrix} 1 & 0 & 3 & 3 & 2 \\ 0 & 3 & 0 & 2 & 3 \\ 3 & 0 & 2 & 0 & 3 \\ 3 & 2 & 0 & 3 & 0 \\ 2 & 3 & 3 & 0 & 1 \end{pmatrix}$$

Answer:

$$= \frac{\begin{vmatrix} -89 & -11 \\ -11 & -89 \end{vmatrix}}{10} = \frac{7800}{10} = 780$$

Remember, when finding a determinant and expanding by cofactors, the most efficient method is to expand by the row or column that has the most zeros in it that you can find. Also, be very meticulous when using any method to find a determinant of a large matrix. It only takes one mistake to throw it all off.

FINAL EXAM

Part I—Multiple Choice
Time Limit—3 Hours

1. Determine which of the following are solutions of the equation $3x^3 - 11x^2 - 6x + 8 = 0$
   a) 0
   b) -1
   c) 4
   d) all of these
   e) None of these

Answer: d

Using synthetic division, you will find that a, b, and c are all roots. So d is the correct answer.
   Another simple way to solve this problem would be just to put each of a, b, and c into the equation to see which works. This avoids synthetic division altogether.

2. Given $f(x) = \sqrt{2x-1}$ find $f^{-1}(x)$

   a) $\sqrt{2x-1}$, $y \geq \dfrac{1}{2}$

   b) $x^2 + 1$, $x \geq 0$

   c) $\dfrac{x^2 + 1}{2}$, $x \geq 0$

   d) $\dfrac{1}{\sqrt{2x-1}}$, $x \geq \dfrac{1}{2}$

   e) none of these

Answer: a

$$x = \sqrt{2y-1} \to x^2 = 2y-1 \to 2y = x^2+1 \to y = \frac{x^2+1}{2} \to f^{-1}(x) = \frac{x^2+1}{2}$$

(which is a good function).

The range of $f(x)$ is $x \geq 0$, so the domain of $f^{-1}(x)$ is $x \geq 0$.

You need to have a good understanding of functions, their domain, and their range, or a question like this might throw you.

3. Determine the amount of money that should be invested at a rate of 6.5% compounded monthly to produce a final balance of $15,000 in 20 years.

  a) $4,102.34
  b) $5,216.07
  c) $2,458.83
  d) $14,056.14
  e) None of these

Answer: a

$$15000 = P\left(1+\frac{.065}{12}\right)^{12*20} \to 15000 = P\left(\frac{12.065}{12}\right)^{240} \to 15000 = 3.656447P \to P = 4102.34$$

4. Find the domain of the function

$$f(x) = \frac{1}{\sqrt{x^2-1}}$$

  a) $(-\infty,-1),(-1,1),(1,\infty)$
  b) $(-\infty,0),(0,\infty)$
  c) $(-\infty,\infty)$
  d) $(-\infty,-1),(1,\infty)$

Answer: d

The x values that make the function undefined are those that are not contained in the domain. So look for these.

First, $x^2 - 1 \geq 0$, since you can only take the square root of a nonnegative number (and still get a real number as an answer).

So $x^2 \geq 1 \Rightarrow x \geq 1$ and $x \leq -1$.

Also $\sqrt{x^2-1} \neq 0$ (or our function would be undefined).

So $x^2 - 1 \neq 0 \Rightarrow x \neq 1$ and $x \neq -1$.

Therefore, the answer is d.

5. Given $f(x) = 2x^2 + 1$, $g(x) = x - 2$ find $(f \circ g)(x)$

   a) $x^2 - 7$

   b) $2x^2 + x - 1$

   c) $2x^2 - 1$

   d) $2x^2 - 8x + 9$

   e) None of these

Answer: d

> $(f \circ g)(x) = 2(x - 2)^2 + 1 = 2(x^2 - 4x + 4) + 1 = 2x^2 - 8x + 8 + 1 = 2x^2 - 8x + 9$
>
> So the answer is d.
>
> Remember that $g(x)$ is what gets put into $f(x)$ instead of x in this case.

6. Find the determinant of the matrix

$$\begin{pmatrix} 0 & -1 & 2 \\ 3 & 5 & 0 \\ 1 & -1 & 3 \end{pmatrix}$$

   a) 25

   b) −25

   c) 7

   d) −7

   e) None of these

Answer: d

> By cofactor expansion along the first row, we get
>
> $$\begin{vmatrix} 3 & 0 \\ 1 & 3 \end{vmatrix} + 2\begin{vmatrix} 3 & 5 \\ 1 & -1 \end{vmatrix} = 9 + 2 * -8 = 9 - 16 = -7$$
>
> So the answer is d.
>
> Be sure to pick a row or column that has as many zeros as possible in it.

7. Find the eighth term of the sequence $5, 13, 21, 29, \ldots$

   a) 61

   b) 53

   c) 69

   d) 56

   e) None of these

Answer: a

> Since it is an arithmetic sequence whose difference is 8, we have
>
> $$a^8 = a^1 + (n - 1)d = 5 + (8 - 1)8 = 5 + 56 = 61$$
>
> So the answer is a.
>
> It should be second nature, when dealing with sequences, to check and see if they are arithmetic or geometric. Arithmetic has the same difference between terms, geometric has the same ratio.

8. Given the system of equations

$$3x + y - z = -11$$
$$x - y - z = -5$$
$$x + 2y + 3z = 3$$

the correct value for x is

a) -3

b) 3

c) -2

d) 2

e) None of these

Answer: a

Use Cramer's rule. (Since we only want to know one of the variables, this involves less work.)

$$D_x = \begin{vmatrix} -11 & 1 & -1 \\ -5 & -1 & -1 \\ 3 & 2 & 3 \end{vmatrix} = \frac{\begin{vmatrix} -11 & 1 \\ -5 & -1 \\ -5 & -1 \\ 3 & 2 \end{vmatrix} \begin{vmatrix} 1 & -1 \\ -1 & -1 \\ -1 & -1 \\ 2 & 3 \end{vmatrix}}{-1} = -\begin{vmatrix} 16 & -2 \\ -7 & -1 \end{vmatrix} = -16 - 14 = -30$$

$$D = \begin{vmatrix} 3 & 1 & -1 \\ 1 & -1 & -1 \\ 1 & 2 & 3 \end{vmatrix} = \frac{\begin{vmatrix} 3 & 1 \\ 1 & -1 \\ 1 & -1 \\ 1 & 2 \end{vmatrix} \begin{vmatrix} 1 & -1 \\ -1 & -1 \\ -1 & -1 \\ 2 & 3 \end{vmatrix}}{-1} = -\begin{vmatrix} -4 & -2 \\ 3 & -1 \end{vmatrix} = 4 + 6 = 10$$

$$x = \frac{D_x}{D} = \frac{-30}{10} = -3$$

So the answer is a.
  Note that you can also use other methods to solve this problem.

9. You need to figure out how many license plates can be made for your state. The stipulations are:

　　1) You cannot use the letters x and z.

　　2) Each license plate consists of 3 numbers followed by 3 letters.

　　3) No personalized license plates are allowed.

　　4) You cannot use the same number twice.

How many different plates can be made?

a) 8,743,680

b) 9,953,280

c) 11,232,000

d) 17,576,000

e) None of these

Answer: b

The first number can be any one of the 10 digits 0–9. The second number can be any one except for the one use in the first number, so 9 possible digits. Likewise, the third can be any except for the two digits already used, so 8 possible digits.

Any of the three letters can be any of the 24 remaining letters of the alphabet after removing x and z, so 24 possible letters.

Number of plates is $10 \times 9 \times 8 \times 24 \times 24 \times 24 = 9{,}953{,}280$. So the answer is b.

It needs to be understood, in this case, the a digit is not affected by the other digits, except when determining which digit it can be.

10. Use the binomial theorem to expand $(x - 4y)^4$
    a) $x^4 + 16x^3y + 96x^2y^2 + 256xy^3 + 256y^4$
    b) $x^4 - 16x^3y + 96x^2y^2 - 256xy^3 + 256y^4$
    c) $x^4 - 4x^3y + 6x^2y^2 + 4xy^3 + 4y^4$
    d) $x^4 + 4x^3y + 6x^2y^2 + 4xy^3 + 4y^4$

Answer: b

$$(a+b)^4 = a^4 + 4a^3b + 6a^2b^2 + 4ab^3 + b^4$$
$$\text{So. } (x+(-4y))^4 = (x)^4 + 4(x)^3(-4y) + 6(x)^2(-4y)^2 + 4(x)(-4y)^3 + (-4y)^4$$
$$= x^4 + 4x^3(-4)y + 6x^2(-4)^2y^2 + 4x(-4)^3y^3 + (-4)^4y^4$$
$$= x^4 - 16x^3y + 96x^2y^2 - 256xy^3 + 256y^4$$

So the answer is b.
   Just for guessing's sake, one could recognize that the last term of the expansion would be the last term to the fourth power, which eliminates c and d as possible answers.

11. h varies jointly with the square of x and the cube root of y. If $h =$ when $x = 2$ and
    $y = 8$, find the constant of proportionality.
    a)
    b) 4
    c) $\frac{1}{32}$
    d) 64
    e) None of these

Answer: e

$$h = k\frac{x^2}{\sqrt[3]{y}} \rightarrow \frac{1}{4} = k\frac{2^2}{\sqrt[3]{8}} \rightarrow \frac{1}{4} = k\frac{4}{2} \rightarrow 2 = 16k \rightarrow k = \frac{1}{8}$$

So the answer is e.

12. Use the change of base formula to identify the expression that is equivalent to $\log_3 5$.

a) $\dfrac{\log 5}{\log 3}$

b) $\dfrac{\ln 3}{\ln 5}$

c) $5 \ln 3$

d) $\log \dfrac{5}{3}$

e) none of these

Answer: a

> The change of base formula says $\log_b a = \dfrac{\log_c a}{\log_c b}$
>
> So, $\log_3 5 = \dfrac{\log_c 5}{\log_c 3}$
>
> So the answer is a (since $c = 10$ is acceptable).
> A good understanding of how logs and exponents are defined will greatly ease many frustrations on algebra final exams. This is an area that most students ignore sufficiently to hurt their grades.

13. Find the sum of the first 15 terms of the sequence defined by

$$a_n = 3\left(\frac{5}{4}\right)^n$$

a) 329.06

b) 411.33

c) 322.61

d) 271.15

e) None of these

Answer: b

> $$S_n = \frac{a_1(1-r^n)}{1-r} \rightarrow S_{15} = \frac{3\left(\frac{5}{4}\right)\left(1-\left(\frac{5}{4}\right)^{15}\right)}{1-\left(\frac{5}{4}\right)} = \frac{\left(\frac{15}{4}\right)(-27.422)}{-\frac{1}{4}} = -15(-27.422) = 411.33$$
>
> So the answer is b.

14. Find the equation of the line that is perpendicular to $2x + 3y = 12$ but has the same y-intercept.

   a) $2x + 3y = 8$
   b) $2x - 3y = 12$
   c) $2x + 3y = 12$
   d) $3x - 2y = -8$
   e) None of these

Answer: d

> The y-intercept of this line is 4 and its slope is $-\frac{2}{3}$. (Find it by putting the line in slope-intercept form.)
> So the perpendicular line must have a slope of
>
> $$\frac{-1}{-\frac{2}{3}} = \frac{3}{2}$$
>
> with a y-intercept of 4. Putting that into the slope-intercept form gives us
>
> $$y = \frac{3}{2}x + 4 \rightarrow 2y = 3x + 8 \rightarrow 3x - 2y = -8$$
>
> So the answer is d.

15. Find the standard equation of the circle:
$$x^2 + y^2 - 2x + 8y - 20 = 0$$
   a) $(x-1)^2 + (y+4)^2 = 20$
   b) $(x-1)^2 + (y+4)^2 = 3$
   c) $(x-2)^2 + (y+8)^2 = 48$
   d) $(x-1)^2 + (y+4)^2 = 37$
   e) None of these

Answer: d

> Complete the square for both x and y and put into standard form.
> $x^2 - 2x + y^2 + 8y - 20 = 0 \rightarrow (x-1)^2 - 1 + (y+4)^2 - 16 - 20 = 0 \rightarrow (x-1)^2 + (y+4)^2 = 37$.
> Completing the square will be used in many places in algebra. It is important to know it well (as well as why it works).

16. Simplify: $5 \ln \sqrt[5]{e^3 x}$

   a) $3e + \ln x$

   b) $3e + 5 \ln \dfrac{x}{5}$

   c) $3 + 5 \ln \dfrac{x}{5}$

   d) $3 + \ln x$

   e) none of these

Answer: d

$$5 \ln \sqrt[5]{e^3 x} = 5 \ln \left(e^3 x\right)^{\frac{1}{5}} = \frac{5}{5} \ln \left(e^3 x\right) = \ln e^3 + \ln x = 3 \ln e + \ln x = 3 + \ln x$$

So the answer is d.
   A good understanding of how logs and exponents are defined will greatly ease many frustrations on algebra final exams. This is an area that most students tend to neglect, with bad effect on grades.

Part II—Workout Problems

1. Given $C = \begin{pmatrix} 1 & 1 & 0 \\ 3 & 1 & 2 \\ -1 & 1 & -1 \end{pmatrix}$, find C^{-1}

Answer:

   Find det C and use the adjoint method of finding the inverse.

$$\det C = \begin{vmatrix} 1 & 1 & 0 \\ 3 & 1 & 2 \\ -1 & 1 & -1 \end{vmatrix} = \begin{vmatrix} 1 & 2 \\ 1 & -1 \end{vmatrix} - \begin{vmatrix} 3 & 2 \\ -1 & -1 \end{vmatrix} = -3 - (-1) = -3 + 1 = -2$$

$$C^{-1} = \frac{\text{adj } C}{\det C} = -\frac{1}{2} \begin{pmatrix} \begin{vmatrix} 1 & 2 \\ 1 & -1 \end{vmatrix} & -\begin{vmatrix} 1 & 0 \\ 1 & -1 \end{vmatrix} & \begin{vmatrix} 1 & 0 \\ 1 & 2 \end{vmatrix} \\ -\begin{vmatrix} 3 & 2 \\ -1 & -1 \end{vmatrix} & \begin{vmatrix} 1 & 0 \\ -1 & -1 \end{vmatrix} & -\begin{vmatrix} 1 & 0 \\ 3 & 2 \end{vmatrix} \\ \begin{vmatrix} 3 & 1 \\ -1 & 1 \end{vmatrix} & -\begin{vmatrix} 1 & 1 \\ -1 & 1 \end{vmatrix} & \begin{vmatrix} 1 & 1 \\ 3 & 1 \end{vmatrix} \end{pmatrix} = -\frac{1}{2} \begin{pmatrix} -3 & 1 & 2 \\ 1 & -1 & -2 \\ 4 & -2 & -2 \end{pmatrix}$$

$$= \begin{pmatrix} \frac{3}{2} & -\frac{1}{2} & -1 \\ -\frac{1}{2} & \frac{1}{2} & 1 \\ -2 & 1 & 1 \end{pmatrix}$$

   Be meticulous when doing these types of calculations. Go slowly and be sure of your work. We all have the tendency to make simple arithmetic errors.

2. Graph the function $f(x) = \dfrac{2x}{x^2 + x - 2}$, labeling all intercepts and asymptotes.

Answer:

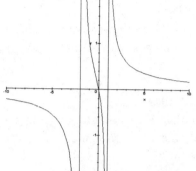

> Remember, the *x*-intercepts are found by finding the zeros of the numerator and the vertical asymptotes are found by finding the zeros of the denominator. Additionally, only one test point needs to be graphed for each section of the graph between *x*-intercepts, vertical asymptotes, and plus or minus infinity.
>
> Note that a horizontal asymptote exists only if the degree of the denominator is less than or equal to the degree of the numerator: If less than, $y = 0$ is the horizontal asymptote. If equal to, $y = a/b$ is the horizontal asymptote where *a* is the leading coefficient of the numerator and *b* is the leading coefficient of the denominator.

3. At what interest rate would a deposit of $30,000 grow to $2,540,689 in 40 years with continuous compounding?

Answer: 11.1 percent.

$$A = Pe^{rt} \rightarrow 2540689 = 30000e^{40t} \rightarrow 84.6896 \approx e^{40t} \rightarrow \ln 84.6896 \approx \ln e^{40t} \rightarrow$$
$$4.439 \approx 40t \ln e \rightarrow 4.439 \approx 40t \rightarrow t \approx \frac{4.439}{40} \approx .111$$

> Make sure you fill in all the information in the equation. There is only one variable to be solved for. If you leave out a piece of information, it could get complicated or you could become confused.

4. Use the rational zero theorem, Descartes' rule of signs, and the theorem on bounds as aids in finding all zeros for the polynomial
$$p(x) = 2x^4 - 11x^3 - 6x^2 + 64x + 32$$
   a) List possible rational zeros.
   b) List the number of positive and negative zeros using Descartes' rule of signs.
   c) List ALL of the zeros of the polynomial.
   d) Graph the polynomial.

Answers:
   a) ±32, ±16, ±8, ±4, ±2, ±1, ±
   b) Positive Zeros: 2,0 (there are two sign changes in $p(x)$)
      Negative Zeros: 2,0 (there are two sign changes in $p(-x)$)
   c) 4,4 – ,2
   d)

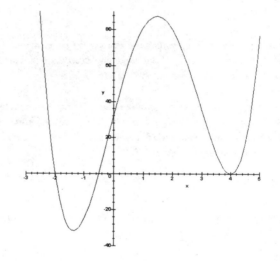

5. If $A = \begin{pmatrix} 1 & 2 & -1 \\ 0 & 1 & 1 \end{pmatrix}$, $B = \begin{pmatrix} 1 & 3 \\ -2 & 4 \end{pmatrix}$, $C = \begin{pmatrix} 2 & -1 \\ 0 & 4 \\ 1 & 3 \end{pmatrix}$, find $B - 2AC$.

Answer:
$$B - 2AC = \begin{pmatrix} 1 & 3 \\ -2 & 4 \end{pmatrix} - 2\begin{pmatrix} 1 & 2 & -1 \\ 0 & 1 & 1 \end{pmatrix}\begin{pmatrix} 2 & -1 \\ 0 & 4 \\ 1 & 3 \end{pmatrix} = \begin{pmatrix} 1 & 3 \\ -2 & 4 \end{pmatrix} - 2\begin{pmatrix} 1 & 4 \\ 1 & 7 \end{pmatrix} = \begin{pmatrix} 1 & 3 \\ -2 & 4 \end{pmatrix} - \begin{pmatrix} 2 & 8 \\ 2 & 14 \end{pmatrix} = \begin{pmatrix} -1 & -5 \\ -4 & -10 \end{pmatrix}$$

Remember your order of operations!

6. Solve the system by the method of substitution:

$$x^2 + 2y = 6$$
$$2x + y = 3$$

Answer:

$2x + y = 3 \Rightarrow y = 3 - 2x \Rightarrow x^2 + 2(3 - 2x) = 6 \Rightarrow x^2 - 4x = 0 \Rightarrow x(x - 4) = 0 \Rightarrow x = 0, 4$

$y = 3 - 2(0) = 3, y = 3 - 2(4) = 3 - 8 = -5$

So $(0,3)$ and $(4,-5)$ are solutions.

As a rule, you should go back and check to make sure your answers work:

$$0^2 + 2 \times 3 = 0 + 6 = 6$$
$$2 \times 0 + 3 = 0 + 3 = 3$$

This obviously works, and so does:

$$4^2 + 2 \times (-5) = 16 - 10 = 6$$
$$2 \times 4 - 5 = 8 - 5 = 3$$

CHAPTER 14

COAS J112: INTRODUCTION TO COLLEGE MATHEMATICS

Jodi Petersime, Associate Instructor

THIS COURSE IS DESIGNED TO BUILD STUDY SKILLS FOR FUTURE MATHEMATICS COURSES and to prepare students for brief calculus or finite mathematics. The course examinations, three midterms and a final, include problem solving as well as multiple-choice questions that test:

1. Ability to solve equations and inequalities individually and in systems

2. Basic understanding of exponential and logarithmic functions

3. Ability to graph conic sections, solutions for systems of equations, and logarithmic and exponential functions

4. Ability to manipulate rational expressions accurately and efficiently

There is no substitute for sound study habits. It is especially important to:

◆ Do homework assignments every day. A sure way to falter in mathematics courses is to allow yourself to fall behind in assignments.

◆ Read each section and try the assigned problems prior to lecture.

◆ Get help when you don't understand a step, problem, or topic. If you wait until the exam, it's too late!

◆ Ask questions in class. Chances are, others have the same question. Don't hesitate to ask.

◆ Take full advantage of all the help that is offered, including office hours and free help sessions.

You should make a friend and do your homework with someone. This allows you to check your work and methods; however, prepare for all exams and quizzes alone, on your own. Avoid using a calculator to prepare for an exam if the use of a calculator will not be allowed in the actual exam.

Do practice problems. Math is not a spectator course. You cannot learn it through osmosis.

Here are some more study hints that will help you throughout a mathematics course and will help prepare you for the exams:

◆ When you begin a new mathematical subject area, first survey the material to get a general idea of the topic. Don't let yourself get hung up on details at the outset.

◆ Take special note of any new terms or symbols. Be certain that you understand these. Don't skip them or shunt them aside.

◆ Read! Mathematics is not all numbers, but often involves careful reading of instructions and explanations before working a problem.

◆ As you study, focus your attention on *why* and *how*. *Why* is the reasoning that governs the solution of a problem, and *how* is the method or process needed to solve a given problem.

◆ As you study, analyze examples. Pay special attention to the order or sequence of steps through which example problems are solved.

◆ Orderly solution procedure is important, but also consider why each step is necessary. You should not only be able to solve problems, but also be able to explain your reason for doing the work in a certain way.

◆ Learn to use your mistakes. They are valuable aids to learning. Don't turn from errors in frustration, but go back through the steps to see how the error developed. This is actually one of the most effective methods of study for mathematics.

Let's return briefly to the subject of reading in mathematics. Few of the problems in this course are word problems; however, you still have to read the mathematical expressions given. As you read them, ask yourself three questions:

1. What is given? What are the *facts* of the problem?

2. What is unknown? What is the answer to be found?

3. How do I proceed? What methods or steps are required to solve the problem?

Your ability to read mathematical expressions will improve rapidly with practice, as you become familiar and comfortable with more terms and symbols. Learn to pay especially close attention to relationships. How does one fact, idea, or mathematical expression lead to another?

Examinations are very important in this course. While homework scores show how hard a student is working, exams demonstrate the level of competency the student has mastered. Together, the three midterm exams make up 60 percent of the student's final grade.

Look through the entire test before you begin to work. First, complete the problems that take little time and are easy for you. Then do the more lengthy problems that you are familiar with. Last, work the problems that you have few ideas for and require significant amounts of time. Never leave any problem blank. Always write down everything you know. Instructors cannot give partial credit for something they cannot see. Show all of your work, and do not take shortcuts. Remember, even if an error is made, partial credit can be salvaged for correct work present.

A student who has mastered the material should be able to work all of the problems within the 55 minutes allowed. While all of the questions are important, 3, 5, 6, 7, 9, and 11 cover the major concepts. Section by section, it is most important to focus on sections II, III, and IV. Together, they account for three-fourths of the grade.

Section I
Solve.

1. $a^2 = 2(a + 1)$
Answer:

$$a^2 = 2(a + 1)$$

Distribute.

$$a^2 = 2a + 2$$

$$\begin{array}{r} -2a \quad -2a \\ a^2 - 2a = 2 \end{array}$$

Complete the square.

$$a^2 - 2a + 1 = 2 + 1$$

A common mistake is to forget to add 1 to both sides.

$$(a - 1)^2 = 3$$

$$(a - 1) = \pm\sqrt{3}$$

Forgetting the $\pm$ is a common error.

$$a = 1 \pm \sqrt{3}$$

Therefore $a = 1+\sqrt{3}$ or $a = 1-\sqrt{2}$

This problem could also have been solved by using the quadratic formula. Credit would be given for either method, provided that adequate documentation is displayed. Students must supply two solutions for *a*.

2. $(2 + 4x)^2 = 8$

Answer:

$$(2+4x)^2 = 8$$
$$2+4x = \pm \sqrt{8}$$
$$2+4x = \pm 2\sqrt{2}$$

CASE 1:

$$2+4x = 2\sqrt{2}$$
$$\underline{-2 \qquad -2}$$
$$\frac{4x}{4} = \frac{2+2\sqrt{2}}{4}$$
$$x = \frac{-2+2\sqrt{2}}{4}$$
$$x = \frac{2(-1+\sqrt{2})}{4}$$
$$x = \frac{-1+\sqrt{2}}{2}$$

CASE 2:

$$2+4x = -2\sqrt{2}$$
$$\underline{-2 \qquad -2}$$
$$\frac{4x}{4} = \frac{-2-2\sqrt{2}}{4}$$
$$x = \frac{-2-2\sqrt{2}}{4}$$
$$x = \frac{2(-1-\sqrt{2})}{4}$$
$$x = \frac{-1-\sqrt{2}}{2}$$

Splitting the solution into two parts is a way that you can check your work; however, the method used to solve problem 1 will also work here, too. Notice that Case 1: $x = \dfrac{-1+\sqrt{2}}{2}$ and Case 2: $x = \dfrac{-1-\sqrt{2}}{2}$ are actually the same as $x = \dfrac{-1\pm\sqrt{2}}{2}$.

3. $a^4 - 5a^2 + 4 = 0$

Answer:

$$a^4 - 5a^2 + 4 = (a^2)^2 - 5(a^2)^1 + 4$$

let $m = a^2$

$$m^2 - 5m + 4 = 0$$

$$(m - 1)(m - 4) = 0$$

$$m - 1 = 0 \text{ or } m - 4 = 0$$

substitute a^2 back in for m

$$a^2 - 1 = 0 \text{ or } a^2 - 4 = 0$$

$$a = \pm 1 \text{ or } a = \pm 2$$

Therefore a = 1, -1, 2 or -2

> Avoid the common mistake of leaving answer in terms of *m* (that is, *m* = 1 and *m* = 4). The question asks for *a*, not *m*. Many students are able to do this problem without substituting *m*. This approach is fine, as long as adequate documentation is provided.

4. $3x^{2/3} - 8x^{1/3} + 4 = 0$

Answer :

$$3\left(x^{1/3}\right)^2 - 8\left(x^{1/3}\right)^1 + 4 = 0$$

let $m = x^{1/3}$

$$3m^2 - 8m + 4 = 0$$

$$(3m - 2)(m - 2) = 0$$

$$3m - 2 = 0 \text{ or } m - 2 = 0$$

substitute back in for m

$$3x^{1/3} - 2 = 0 \text{ or } x^{1/3} - 2 = 0$$

$$\phantom{3x^{1/3}} + 2 \ +2 \qquad +2 \ +2$$

$$\frac{3x^{1/3}}{3} = \frac{2}{3} \text{ or } x^{1/3} = 2$$

$$\left(x^{1/3}\right)^3 = \left(\frac{2}{3}\right)^3 \text{ or } \left(x^{1/3}\right)^3 = (2)^3$$

$$x = \frac{8}{27} \text{ or } x = 8$$

> Again, many students can solve this problem without substituting. This is fine. Students must factor and then solve—with or without substitution.

5. $x^{-2} + 2x^{-1} - 3 = 0$

Answer:

$$(x^{-1})^2 + 2(x^{-1}) - 3 = 0$$

let $m = x^{-1}$

$$m^2 + 2m - 3 = 0$$

$$(m - 1)(m + 3) = 0$$

$$m - 1 = 0 \text{ or } m + 3 = 0$$

substitute back in form

$$x^{-1} - 1 = 0 \text{ or } x^{-1} + 3 = 0$$

$$\begin{array}{cc} +1 \quad +1 & -3 \quad -3 \end{array}$$

$$x^{-1} = 1 \qquad x^{-1} = -3$$

$$(x^{-1})^{-1} = (1)^{-1} \quad (x^{-1})^{-1} = (-3)^{-1}$$

$$x = 1 \quad \text{or} \quad x = \frac{1}{-3} = -\frac{1}{3}$$

Section II

Solve each and express in interval notation.

6. $x^2 + 4x \geq 5$

Answer:

$$x^2 + 4x - 5 \geq 0$$

Zero on one side. Then factor and solve.

$$(x^2 - 1)(x + 5) = 0$$

$$x - 1 = 0 \text{ or } x + 5 = 0$$

$$x = 1 \text{ or } x = -5$$

$$\begin{array}{cccc} (+) & & (+) \\ \leftarrow \overline{}_{-6} \ \overline{}_{-5} \ \overline{}_{-4} \ \overline{}_{-3} \ \overline{}_{-2} \ \overline{}_{-1} \ \overline{}_{0} \ \overline{}_{1} \ \overline{}_{2} \ \overline{}_{3} \rightarrow \end{array}$$ graph points

Choose one point from each interval to test (+) or (–).

Test points

x	$x^2 + 4x - 5$
-6	Plug in $(-6)^2 + 4(-6) - 5 = 36 - 24 - 5 = (+)$
-1	Plug in $(-1)^2 + 4(-1) - 5 = +1 - 4 - 5 = (-)$
2	Plug in $(2)^2 + 4(2) - 5 = 4 + 8 - 5 = (+)$

We need the solution to be greater than zero, or positive; therefore, our solution is $(-\infty, -5] \cup [1, \infty)$.

> I am looking to see that students factor and solve correctly, choose and test three points, analyze their solutions correctly, and understand interval notation. Students must remember $-\infty$ and $+\infty$ always use parentheses. Brackets are used for numbers that can be included in the solution.

7. $\dfrac{(x-1)(x+5)}{x+2} < 0$

Answer:

The rational expression is undefined when #/0 or $x + 2 = 0$ or $x = -2$.

The rational expression is equal to zero when 0/# or $(x-1)(x+5) = 0$ or when

$$x - 1 = 0 \quad x + 5 = 0$$
$$x = 1 \text{ or } x = -5$$

Test points

x	$\dfrac{(x-1)(x+5)}{x+2}$
-6	$\dfrac{(-)(-)}{(-)} = (-)$
-4	$\dfrac{(-)(+)}{(-)} = (=)$
-1	$\dfrac{(-)(+)}{(+)} = (-)$
2	$\dfrac{(+)(+)}{(+)} = (+)$

$$\frac{(-6-1)(-6+5)}{-6+2} = \frac{(-)(-)}{(-)} = (-)$$

We are looking for <0 or negative. Therefore, our solution is $(-\infty, -5) \cup (-2, 1)$.

You must:
   1. Find the values of x that make the rational expression undefined (#/0).
   2. Find the values of x that make the rational expression equal to 0 (0/#).
   3. Test points within each interval.
   4. Analyze the solution.
   5. Express it in interval notation.

Section III

Solve each system.

8. $x + 2y = 3$ $(E1)$
   $3x + y = -1$ $(E2)$

Answer:

Solving by substitution:

$$(E1)\ x + 2y = 3$$
$$-2y - 2y$$
$$x = 3 - 2y$$

Check your solution! $(-1, 2)$

$$(E1)\ -1 + 2(2) \overset{?}{=} 3$$
$$-1 + 4 = 3\ \checkmark$$

$$(E2)\ 3(3 - 2y) + y = -1$$
$$9 - 6y + y = -1$$
$$9 - 5y = -1$$
$$-9 \quad -9$$
$$-5y = -10$$
$$y = 2$$

Check solution $(E2)$:

$$3(-1) + 2 \overset{?}{=} -1$$
$$-3 + 2 = -1\ \checkmark$$

$$(E1)\ x = 3 - 2(2)$$
$$x = 3 - 4$$
$$x = -1$$

Therefore, our solution is $(-1, 2)$.

A common mistake: $3 \times 3 - 2y + y = -1$

$9 - 2y + y = -1$

Make certain to distribute the 3 to $-2y$ properly.

There are many different ways to begin this problem using substitution. The goal is to solve one of the equations for a variable and substitute it into the other equation. In addition to solving by substitution, the problem can be solved using elimination. There are also many different approaches using elimination.

A good solution is a well-documented one. The solution should be written neatly, in an orderly fashion, as if it were a story. Students are crazy if they do not check their answers. A solution can be checked by substituting the numbers in for the variables, making sure this leads to a true statement. Both equations must be checked. A solution for one equation is not always a solution for the other. It is also important to check answers because many arithmetic errors are possible.

9. $x + y - 2z = -6$ $(E1)$

    $2y + z = 2$ $(E2)$

    $x + z = 0$ $(E3)$

Answer:

Solving by elimination:

$$
\begin{array}{ll}
x + y - 2z = -6 & (E1) \\
\underline{-x \quad\quad - z = 0} & \underline{-1(E3)} \\
\quad\quad y - 3z = -6 & (E1) + -1(E3) = (E4)
\end{array}
$$

$$
\begin{array}{ll}
2y + \;\; z = 2 & (E2) \\
\underline{-2y + 6z = 12} & \underline{-2\,(E4)} \\
\quad\quad\quad 7z = 14 & (E2) + -2(E4) \\
\quad\quad\quad\;\; z = 2 &
\end{array}
$$

$$
\begin{array}{ll}
x + 2 = 0 & (E3) \text{ Substitute 2 in for } z \\
\quad\; x = -2 &
\end{array}
$$

$$
\begin{array}{ll}
-2 + y - 2(2) = -6 & (E1) \text{ Substitute 2 for } z \text{ and } -2 \text{ for } x \\
-2 + y - 4 = -6 & \\
\quad\; y - 6 = -6 & \\
\quad\quad\;\; y = 0 &
\end{array}
$$

Therefore, our solution is $(-2,0,2)$.

Check step:

$(E1)$ $-2 + 0 - 2(2) = -6$

              $-2 - 4 = -6$✔

$(E2)$ $2(0) + 2 = 2$

                  $2 = 2$✔

$(E3)$ $-2 + 2 = 0$✔

To use the elimination method:
1. Combine two equations to eliminate one of the variables. (E1) and a multiple (–1) of (E3) we combine to give a new (E4).
2. Next, combine two different equations to eliminate the same variable, again giving (E5). (E2) itself already had the x's eliminated, so we were able to use (E2) instead of working for a new (E5).
3. Combine the two preceding equations to solve for a variable. In this case, we eliminated the y's and solved for z.
4. Substitute z back into (E4) or (E5) (E2, in our case) to solve for y.
5. Substitute y and z to solve for x.
6. Finally, check your solution.

There are many different ways one could use elimination to solve this problem. The general method for any of the approaches should be similar. The most important thing to remember is to keep work neat, orderly, and well documented. There is no chance for partial credit if the instructor cannot easily follow your work. It is also important to keep x's, y's, and z's lined up vertically. Remember, it is wrong to add $6x + 5y$ and get $11xy$. The labels $E\#$ + 2 ($E\#$) are very helpful to the student and instructor for finding mistakes and giving partial credit.

This problem could also have been solved using substitution.

Section IV

Use row reduction of matrices to solve.

10. $x + 3y + 4z = 3$ (E1)
   $-x + y - 3z = 4$ (E2)
   $x - 3y + z = -6$ (E3)

Answer:

Augmented matrix

$$
\begin{array}{c}
E1 \\ E2 \\ E3
\end{array}
\left[\begin{array}{ccc|c}
1 & 3 & 4 & 3 \\
-1 & 1 & -3 & 4 \\
1 & -3 & 1 & -6
\end{array}\right]
\begin{array}{c}(E2)+(E3)\to(E3)\\ \Rightarrow\end{array}
\left[\begin{array}{ccc|c}
1 & 3 & 4 & 3 \\
-1 & 1 & -3 & 4 \\
0 & -2 & -2 & -2
\end{array}\right]
\begin{array}{c}-\frac{1}{2}(E3)\to(E3)\\ \Rightarrow\end{array}
\left[\begin{array}{ccc|c}
1 & 3 & 4 & 3 \\
-1 & 1 & -3 & 4 \\
0 & 1 & 1 & 1
\end{array}\right]
$$

$$
\begin{array}{c}(E1)+(E2)\to(E2)\\ \Rightarrow\end{array}
\left[\begin{array}{ccc|c}
1 & 3 & 4 & 3 \\
0 & 4 & 1 & 7 \\
0 & 1 & 1 & 1
\end{array}\right]
\begin{array}{c}(E2)+-4(E3)\to(E2)\\ \Rightarrow\end{array}
\left[\begin{array}{ccc|c}
1 & 3 & 4 & 3 \\
0 & 0 & -3 & 3 \\
0 & 1 & 1 & 1
\end{array}\right]
\begin{array}{c}\frac{-1}{3}(E2)\\ \Rightarrow\end{array}
\left[\begin{array}{ccc|c}
1 & 3 & 4 & 3 \\
0 & 0 & 1 & -1 \\
0 & 1 & 1 & 1
\end{array}\right]
$$

$$
\begin{array}{c}\text{switch}\\ (E2)\text{ and }(E3)\\ \Rightarrow\end{array}
\left[\begin{array}{ccc|c}
1 & 3 & 4 & 3 \\
0 & 1 & 1 & 1 \\
0 & 0 & 1 & -1
\end{array}\right]
\begin{array}{c}-1(E3)+(E2)\to(E2)\\ \Rightarrow\end{array}
\left[\begin{array}{ccc|c}
1 & 3 & 4 & 3 \\
0 & 1 & 0 & 2 \\
0 & 0 & 1 & -1
\end{array}\right]
\begin{array}{c}-4(E3)+(E1)\to(E1)\\ \Rightarrow\end{array}
\left[\begin{array}{ccc|c}
1 & 3 & 0 & 7 \\
0 & 1 & 0 & 2 \\
0 & 0 & 1 & -1
\end{array}\right]
$$

$$
\begin{array}{c}-3(E2)+(E1)\to(E1)\\ \Rightarrow\end{array}
\left[\begin{array}{ccc|c}
1 & 0 & 0 & 1 \\
0 & 1 & 0 & 2 \\
0 & 0 & 1 & -1
\end{array}\right]
$$

Therefore, our solution is $(1,2,-1)$ or $x = 1$, $y = 2$, and $z = -1$.

Check the solution:

($E1$) $1 + 3(2) + 4(-1) \overset{?}{=} 3$

$\qquad 1 + 6 \quad\ -4 \quad\ \overset{?}{=} 3$

$\qquad\qquad 7 \quad\ -4 \quad\ = 3 ✔$

($E2$) $-1 + 2 - 3(-1) \overset{?}{=} 4$

$\qquad -1 + 2 + 3 \quad\ \overset{?}{=} 4$

$\qquad\quad 1 + \quad 3 \quad\quad = 4 ✔$

($E3$) $1 - 3(2) + -1 \overset{?}{=} -6$

$\qquad 1 - 6 + \quad -1 \quad \overset{?}{=} -6$

$\qquad -5 + \qquad\ -1 \quad = -6 ✔$

The important thing to keep in mind is the goal of the solution. Our goal for three equations and three unknowns is the following:

$$\begin{bmatrix} 1 & 0 & 0 & | & \# \\ 0 & 1 & 0 & | & \# \\ 0 & 0 & 1 & | & \# \end{bmatrix}$$

Therefore, however you wish to proceed, the method must begin by achieving the zeros. Remember, the only tools of manipulation that can be used are the following:

1. Any two rows can be interchanged.
2. Any row can be multiplied, in its entirety, by any nonzero number.
3. A multiple of any row can be added to another.

You must have all equations in variables = constants form before beginning. A row can only be modified by a new row if the row is a component of it. For example:

$$\begin{bmatrix} E1 \\ E2 \\ E3 \end{bmatrix} \quad E3 \text{ cannot be replaced by } -2(E2) \text{ or } \begin{bmatrix} E1 \\ E2 \\ -2(E2) \end{bmatrix}$$

A successful answer is one that is easy to follow and written like a story. The instructor must see $[(E1) + -2(E2)]$ direction beside each change if partial credit is expected.

It is very easy to make mistakes; therefore, it is imperative that you check your solution. Occasionally, if you are unable to find your mistake, it is advantageous to begin the problem again with a different approach.

Section V

Graph the solution of the following system of inequalities:

11. $x \le 4$

$y \ge -1$

$y \le -\dfrac{1}{2}x + 5$

$y - x \le 2$

Answer:

Method—

1. Graph a line.
2. Pick one test point to evaluate. If the test point gives a true solution, shade that side of the line. If it gives a false statement (i.e., $-1 > 0$), shade the opposite side.
3. Repeat for each line.
4. The solution is given by the area covered by all shadings.

Test points

line $x \le 4$	$x = 0$	$0 \le 4$	true
line $y \ge -1$	$y = 0$	$0 \ge -1$	true
line $y \le -x + 5$	$x = 0, y = 0$	$0 \le 0 + 5$	true
line $y - x \le 2$	$x = 0, y = 0$	$0 - 0 \le z$	true

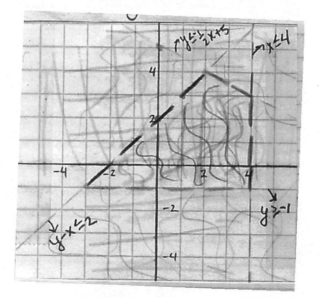

You may find it helpful to use colored pencils for the graph. You must first master graphing a line. Remember, you must use $y = mx + b$ form to get slope and y-intercept information. If you forget this, you can always fall back on plotting points. When finding the common solution, a typical mistake is to use the x- and y-axis as a border of the solution. The only way this should occur is if $x = 0$ or $y = 0$ is a line in the problem (hint: can be replaced with an inequality).

A successful answer is one that generally follows the method in the preceding solution. Test point evaluation should be written down. If a mistake is made, this allows the instructor to locate the error and give partial credit.

FINAL EXAM

The comprehensive final exam counts for 30 percent of the student's final grade. Two hours are allotted, during which all questions should be completed. While all of the exam questions are important, numbers 10 through 17 are especially significant. These give the student an opportunity to demonstrate mastery of graphing conic sections, exponential functions, and logarithmic functions. It is also important to understand how to solve equations and inequalities individually and in systems.

The multiple-choice questions are individually easier, briefer, and less important than the full problems; however, they do make up a considerable portion of the final exam.

Take advantage of the multiple-choice situation by working from the answers—if the solution to a problem is not obvious to you. Often, you can begin by eliminating choices you know are incorrect. Then you can try "plugging in" numbers to replace variables. You can also scan the possible solutions for answers that look *approximately* correct. Try plugging in this solution.

These methods are not the same as guessing (which *is* a useful last resort), but since you don't have to show work in answering multiple-choice questions, they do represent a useful shortcut.

Section I

Solve.

1. $x^{2/3} + x^{1/3} - 2 = 0$

Answer:

$$\left(x^{1/3}\right)^2 + \left(x^{1/3}\right)^1 - 2 = 0$$

let $m = x^{1/3}$

$$m^2 + m - 2 = 0$$

$$(m - 1)(m + 2) = 0$$

replace m with $x^{1/3}$

$$x^{1/3} - 1 = 0 \quad \text{or} \quad x^{1/3} + 2 = 0$$

$$\quad\quad +1 \;\; +1 \quad\quad\quad\quad -2 \; -2$$

$$x^{1/3} = 1 \quad\quad\quad x^{1/3} = -2$$

$$\left(x^{1/3}\right)^3 = (1)^3 \quad \left(x^{1/3}\right)^3 = (-2)^3$$

$$x = 1 \quad\quad\quad\quad x = -8$$

Common mistakes include leaving the answer in terms of *m* instead of *x*.

$$-(1)^3 \neq 3$$

$$-(-2)^3 \neq -6$$

$$-(-2)^3 \neq +8$$

You may be able to skip the substitution step; however, it is important to realize that the factors are

$$\left(x^{1/3} - 1\right)\left(x^{1/3} + 2\right) \text{ not } \left(x^{2/3} - 1\right)\left(x^{2/3} + 2\right).$$

It is important to understand how to factor radical expressions, then solve for the variable.

2. $x^2 \leq 6 - x$

Answer:

$$x^2 \leq 6 - x$$

$$-6 + x - 6 + x$$

$$x^2 + x - 6 \leq 0$$

$$x^2 + x - 6 = 0 \text{ when}$$

$$(x - 2)(x + 3) = 0$$

$$x - 2 = 0 \text{ or } x + 3 = 0$$

$$x = 2 \text{ or } x = -3$$

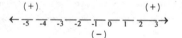

Test points

x	$x^2 + x - 6$
-4	$(-4)^2 + -4 - 6 = (+)$
-1	$(-1)^2 + -1 - 6 = (-)$
3	$(3)^2 + 3 - 6 = (+)$

We are looking for ≤ 0 or negative; therefore, our solution is $[-3, 2]$.

To find a successful solution, you need to do the following:

1. Factor the polynomial to find where it is equal to zero.
2. Test a point within each interval to see if the polynomial is positive or negative within the interval.
3. Analyze the problem: What are we looking for?
4. Graph the solution or put it into interval notation.

You must understand the basic graph of a parabola and why we are able to test only one point within an interval.

Due to the number of arithmetic errors possible, I insist that students test one point from each interval, regardless of the alternating pattern.

3. $x^5 - 6x^3 + 8x = 0$

Answer:

   Notice that there is a common factor of x.

$$x(x^4 - 6x^2 + 8) = 0$$

factoring $x^4 - 6x^2 + 8 = (x^2)^2 - 6(x^2)^1 + 8$

let $m = x^2$

$$m^2 - 6m + 8 = (m - 4)(m - 2)$$

Substitute back in for m

Thus our factorization is

$$x(x^2 - 4)(x^2 - 2) = 0$$

when $x = 0$ or $x^2 - 4 = 0$ or $x^2 - 2 = 0$

$x = 0$ or $x^2 = 4$ or $x^2 = 2$

$x = 0$ or $x = \pm 4$ or $x = \pm\sqrt{2}$

Therefore, our solution is

$$x = 0, 2 - 2, \sqrt{2} \text{ or } -\sqrt{2}$$

Beware of the common error of leaving off the factor of x, thereby leaving the factorization as $(x^2 - 4)(x^2 - 2) = 0$.

   The key is realizing the common factor of x. Once the x is pulled out, you can factor the remaining polynomial. You may be able to factor the remaining polynomial without substitution—which is fine.

4. $\log_3(x + 7) - \log_3(x - 1) = 2$

Answer:

$$\log_3(x + 7) - \log_3(x - 1) = \log_3\left(\frac{x + 7}{x - 1}\right) = 2$$

$$3^2 = \frac{x + 7}{x - 1} \Rightarrow 9 = \frac{x + 7}{x - 1} \Leftrightarrow \frac{9}{1} = \frac{x + 7}{x - 1}$$

Cross multiply:

$$9(x - 1) = 1(x + 7)$$

$$9x - 9 = x + 7$$

$$-x + 9 \quad -x + 9$$

$$8x = 16$$

$$x = 2$$

Note that a logarithmic function can easily be converted to exponential notation. Remember:

$$\log_a b - \log_a c = \log_a \frac{b}{c}$$

$$\log_a b = x \quad \text{or} \quad a^x = b$$

5. $x(2x + 1) = 3$

Answer:

$$2x^2 + x = 3$$
$$2x^2 + x - 3 = 0$$
$$(2x + 3)(x - 1) = 0$$
$$2x + 3 = 0 \quad x - 1 = 0$$
$$x = \frac{-3}{2} \quad \text{or} \quad x = 1$$

This type of problem has been tested previously. You can use factoring, completing the square, or the quadratic formula. The instructor is looking for a demonstration of mastery of factoring through thorough documentation of the method used.

6. $\left(\dfrac{1}{3}\right)^{-6x} = 9^{x+4}$

Answer:

$$\left(\frac{1}{3}\right)^{-6x} = 9^{x+4}$$

$$\left(\left(\frac{1}{3}\right)^{-1}\right)^{6x} = 9^{x+4}$$

$$(3)^{6x} = 9^{x+4}$$
$$(3^2)^{3x} = 9^{x+4}$$
$$3x = x + 4$$
$$2x = 4$$
$$x = 2$$

Begin by getting the same base. There are many different ways to approach this issue, but, regardless of method, all steps must be well documented. The goal of the problem is demonstration of mastery of the following property: $a^x = a^y$ implies $x = y$.

Section II

Solve each system.

7. $x + 2y - z = 1$ $(E1)$
      $-y + 3z = -2$ $(E2)$
    $-x - 2y + 2z = -1$ $(E3)$

Answer:

Using elimination:

$(E1)$ $x + 2y - z = 1$
$(E3)$ $\underline{-x - 2y + 2z = -1}$
$(E1) + (E3)$ $z = 0 = (E4)$
$(E2)\ z = 0$ $-y + 3(0) = -2$
                       $-y = -2$
                       $y = z$
$(E1)\ y = 2\ \ z = 0$ $x + 2(2) - 0 = 1$
                       $x + 4 = 1$
                       $x = -3$

Check your solution:

    $(E1)$ $-3 + 2(z) - 0 \overset{?}{=} 1$
                 $-3 + 4 = 1✔$
    $(E2)$ $-2 + 3(0) = -2✔$
    $(E3) - (-3) - 2(2) + 2(0) \overset{?}{=} -1$
                 $3 - 4 + 0 = -1✔$

There are many ways to begin this problem using elimination. Instructors look for an orderly, neat, and well-documented solution demonstrating mastery of solving three equations with three unknowns.

This problem could also be solved using substitution. No matter the method, all solutions should be checked by plugging the values back into the three equations.

8. $y = x^2 - 4$ $(E1)$
   $y - 2x = -1$ $(E2)$

Answer:

Using substitution
$(E2)$ substituting in for y

$$(x^2 - 4) - 2x = -1$$
$$x^2 - 2x - 4 = -1$$
$$+1 \quad +1$$
$$x^2 - 2x - 3 = 0$$
$$(x + 1)(x - 3) = 0$$
$$x = -1 \quad \text{or} \quad x = 3$$

$$\underline{\text{Case 1 } x = -1} \qquad \underline{\text{Case 2 } x = 3}$$
$$y = (-1)^2 - 4 \qquad\qquad Y = 3^2 - 4$$
$$y = -3 \qquad\qquad\qquad Y = 5$$

Therefore, our two solutions are $(-1,-3)$ and $(3,5)$.

Check step:

$$(-1,3): \qquad -3 = (-1)^2 - 4 = -3 \checkmark \qquad (E1)$$
$$-3 - 2(-1) \overset{?}{=} -1 \qquad\qquad (E2)$$
$$-3 + 2 = -1 \checkmark$$

$$(3,5): \qquad 5 = 3^2 - 4 = 9 - 4 \checkmark \qquad (E1)$$
$$5 - 2(3) \overset{?}{=} -1 \qquad\qquad (E2)$$
$$5 - 6 = -1 \checkmark$$

> This problem is best solved using substitution. Instructors look for a well-documented solution, including the check step.

Section III

Sketch the graph.

9. $x^2 + y^2 + 2x - 8y + 9 = 1$

Answer:

$$x^2 + 2x + y^2 - 8y = -8$$
$$x^2 + 2x + 1 + y - 8y + 16 = -8 + 1 + 16$$

> **Complete the square of x and y.**
>
> $$(x + 1)^2 + (x - 4)^2 = 9$$
>
>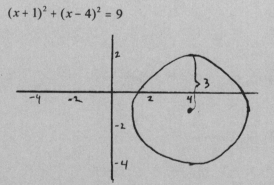
>
> The graph will be a circle centered at $(-1,4)$ with a radius of 3.
>
> The goal of this problem is for students to demonstrate their understanding of graphing a circle. It is necessary to recall how to complete the square and the standard form of an equation for a circle: $(x - h)^2 + (y - k)^2 = r^2$. The circle is centered at (h,k) with radius r.

10. $f(x) = x^2 + 2x - 8 = 0$

Answer:

First find for what values of x is $x^2 + 2x - 8 = 0$

$$(x - 2)(x + 4) = 0$$

$$x = 2 \text{ and } x = -4$$

vertex is located at

$$x = \frac{-b}{2a} \quad \text{or} \quad \frac{-2}{2} = -1$$

$$y \text{ coordinate} = f\left(\frac{-b}{2a}\right) = f(-1) = (-1)^2 + 2(-1) - 8 = 0$$

$$= 1 - 2 - 8 = -9$$

Thus the vertex is $(-1, -9)$

y-intercepts occur when $x = 0$ or $0^2 + 2(0) - 8 = -8$

y-intercept $(0, -8)$

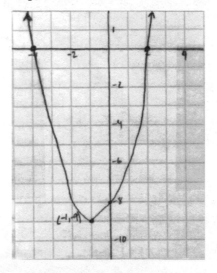

It is important to identify a few main points for $ax^2 + bx + c = 0$.

◆ **Vertex:** $\left(-b/2a, f\left(-b/2a\right)\right) \cdot f\left(-b/2a\right)$

is found by substituting the value $-b/2a$ in for x.

◆ **x-intercepts:** found when $f(x) = 0$. Factor the polynomial.
◆ **y-intercepts:** found when $x = 0$.
If you cannot remember anything, you could still plot points to sketch the graph.

11. $f(x) = 2^{2x-1}$

Answer:

Plot points.

x	$f(x)$
0	$2^{-1} = \frac{1}{2}$
1	$2^{2(1)-1} = 2^1 = 2$
-1	$2^{-2-1} = 2^{-3} = (\frac{1}{2})^3 = 1/8$
2	$2^{2(2)-1} = 2^3 = 8$
$\frac{1}{2}$	$2^{2(\frac{1}{2})-1} = 2^{1-1} = 2^0 = 1$

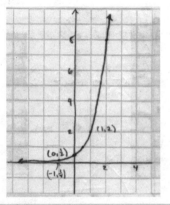

This problem tests your ability to plot points and accurately graph the points. Familiarity with the appearance of an exponential graph is very helpful.

12. $f(x) = \log_2 x$

Answer:

$$f(x) = \log_2 x \text{ means } 2^{f(x)} = x$$

f(x)	x
0	$2^0 = 1$
-1	$2^{-1} = \dfrac{1}{2}$
1	$2^1 = 2$
2	$2^2 = 4$
-2	$2^{-2} = \dfrac{1}{4}$

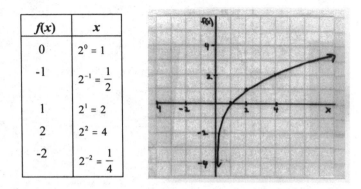

13. $y - \dfrac{-1}{4}(x-1)^2 + 3$

Answer:

Recall the vertex of $y = a(x-h)^2 + k$ is at (h,k).

Thus, the parabola's vertex is at $(1,3)$.

Plot a few more points:

x	y
-1	$\dfrac{-1}{4}(-1-1)^2 + 3 = \dfrac{-1}{4}(4) + 3 = 2$
3	$\dfrac{-1}{4}(3-1)^2 + 3 = 2$
5	$\dfrac{-1}{4}(5-1)^2 + 3 = -4 + 3 = -1$
-3	$\dfrac{-1}{4}(-3-1)^2 + 3 = -4 + 3 = -1$

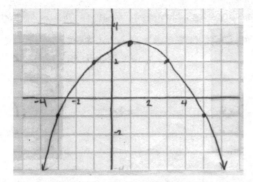

There are many different ways to approach this problem. You could multiply it out, find x-intercepts, y-intercepts, and so on. Most alternatives lead to graph fractions, however, which many students find difficult. If you are careful about the points chosen, you may avoid fractional work.

14. $\frac{1}{3}y + x = -1$

Answer:

$$\frac{1}{3}y + x = -1$$

$$\phantom{\frac{1}{3}y}-x \ -x$$

$$\frac{1}{3}y = -x - 1$$

$$3\left(\frac{1}{3}y\right) = 3(-x-1)$$

$$y = -3x - 3$$

$$\text{slope} = \frac{-3}{1} \quad y\text{-intercept } (0,-3)$$

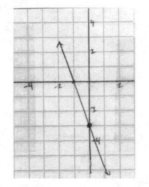

You may either plot points or put the equation into $y = mx + b$ form. m is the slope and $(0,b)$ the y-intercept.

15. $\dfrac{x^2}{25} + \dfrac{y^2}{1} = 1$

Answer:

$$\frac{x^2}{25} + \frac{y^2}{1} = 1$$

Centered at $(0,0)$
5 units (horizontal radius)
1 unit (vertical radius)

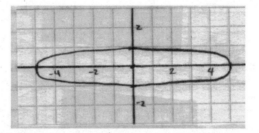

You will need this standard equation for ellipses:

$$\frac{(x - h)^2}{a^2} + \frac{(y - k)^2}{b^2} = 1$$

The ellipse is centered at (h,k). The horizontal radius or axis is a, and the vertical radius or axis is b.

16. $y^2 - 16x^2 = 16$

Answer:

$y^2 - x^2$ tells us we are graphing a hyperbola; therefore, we must put the equation into standard form:

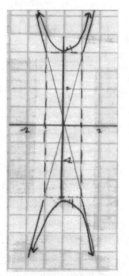

$$\frac{y^2}{a^2} - \frac{y^2}{b^2} = 1 \text{ or}$$

$$\frac{y^2}{a^2} - \frac{x^2}{b^2} = 1$$

$$\frac{y^2}{16} - \frac{16x^2}{16} = \frac{16}{16}$$

$$\frac{y^2}{16} - \frac{x^2}{1} = 1$$

horizontal: 1

vertical: 4

$y^2 > 0 \Rightarrow$ graph opening up and down

This problem requires familiarity with the hyperbola graph and standard form. If the hyperbola is not centered at (0,0), the following would apply:

$$\frac{(y-k)^2}{a^2} - \frac{(x-h)^2}{b^2} = 1 \quad \text{center } (h,k)$$

If $Y^2 < 0$, the graph would open left and right.

Section IV

Graph the solution for the system.

17. $y \geq 0$ (1)

     $y - x \leq 2$ (2)

     $y + \dfrac{3}{2}x - 4 \leq 0$ (3)

Answer:

     First, we must graph each line and shade the proper area.

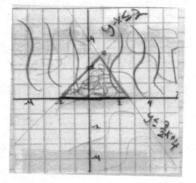

(2) $y - x \leq 2$

       $+ x + x$

     $y \dfrac{2}{1} x + 2$

     $m = \dfrac{1}{1}$ y-intercept $(0,2)$

     Test point

     $(0,0)$: $0 - 0 \leq 2$ true

(3) $y \leq \dfrac{-3}{2} x + 4$

     $m = \dfrac{-3}{2}$ y-intercept $(0,4)$

     Test point

     $(0,0)$: $0 \leq \dfrac{-3}{2}(0) + 4$ true

Begin by graphing each line. To shade the correct side, choose a point not on the line. Test the point. If it gives a true statement, shade that side of the line. If it is false, shade the other side of the line. The solution to the system is the intersection of the shaded areas.

18. $x^2 + \dfrac{y^2}{4} \le 1$ (1)

    $x^2 + y^2 \le 4$ (2)

Answer:

$$x^2 + \frac{y^2}{4} \le 1 \qquad (1)$$

Ellipse: horizontal radius = 1

                vertical radius = 2

Test point: (0,0)

$$0 + \frac{0}{4} \le 1 \quad \text{true}$$

$$x^2 + y^2 \le 4 \; (2)$$

Circle: radius = 2

Test point: (0,0) $0^2 + 0^2 \le 4$ true

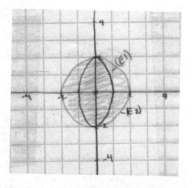

The method for solving this problem is similar to that used in problem 17. It is important to remember what the equations represent:

$$\frac{x^2}{a^2} + \frac{y^2}{b^2} = r^2 \quad \text{circle}$$

$$\frac{x^2}{a^2} - \frac{y^2}{b^2} = 1 \quad \text{hyperbola}$$

$$\frac{x^2}{a^2} + \frac{y^2}{b^2} = 1 \quad \text{ellipse}$$

Section V

Multiple Choice

19. If $\log_r a = m$ and $\log_r b = n$, then $\log_r\left(a^3/b\right) =$

  a. m^3/n
  b. $n - 3m$
  c. $3n/m$
  d. $3m - n$
  e. None of the above

Answer: d

$$\log_r a^3/b = \log_r a^3 - \log_r b = 3\underbrace{\log_r a}_{m} - \underbrace{\log_r b}_{n} = 3m - n$$

20. The inverse of $f(x) = \sqrt{x+1}$ is $f^{-1}(x) =$

  a. $y^2 - 1$
  b. $x^2 - 1$
  c. $x + 1$
  d. x^2
  e. None of the above

Answer: b

$$y = \sqrt{x+1} \quad \text{solve for } x$$
$$y^2 = x + 1$$
$$y^2 - 1 = x \quad \text{switch } x \text{ and } y$$
$$f^{-1}(x) = x^2 - 1$$

21. Simplify

$$\frac{(xy^3 z^{-4})^{-2}}{x^{-3}(yz)^4}$$

  a. $\dfrac{x^2}{y^6 z^8}$

  b. $\dfrac{x^5 y^2}{z^{12}}$

  c. $x^5 y^2 z^4$

  d. $\dfrac{x^2 z^4}{x}$

  e. None of the above

Answer: b

$$\frac{(xy^3z^{-4})^2}{x^{-3}(yz)^4} = \frac{x^2y^{3\cdot2}z^{-4\cdot2}}{x^{-3}y^4z^4} = \frac{x^2x^3y^6}{y^4z^4z^8} = \frac{x^5y^6}{y^4z^{12}} = \frac{x^5y^2}{z^{12}}$$

Remember: $x^{-exponent}$ **means switch position:**

$$\frac{1}{x^{-2}} = \frac{x^2}{1}$$

22. $\log_{16}8 =$

   a.

   b. 2

   c. $\frac{1}{4}$

   d.

   e. None of the above

Answer: c

$$\log_{16}8 = ?\ 16^? = 8\ (16)^? = (2^4)^?8 = 2^3$$

By trial and error:

$$16^{1/2} = 4 \neq 8 \qquad \text{or} \qquad 2^{4\cdot?} = 2^3$$
$$16^2 \neq 8 \qquad \text{or} \qquad 4 \cdot ? = 3$$
$$16^{3/4} = \left(16^{1/4}\right)^3 = 2^3 = 8$$

23. If $f(x) = \log_3 x$ and $g(x) = x^3$, then $f(g(x)) =$

   a. 1

   b. x

   c. $3x$

   d. $3\log_3 x$

   e. None of the above

Answer: d

$$f(g(x)) = f(x^3) = \log_3 x^3 = 3\log_3 x$$

24. The domain of $f(x) = \dfrac{(x+1)(x-1)}{x-3}$ is

   a. All real numbers except $x = 3$

   b. All real numbers except $x = \pm 1$

   c. All real numbers except $x = \pm 1$ and $x = 3$

   d. All real numbers

   e. None of the above

Answer: a

When $x = 3$, the denominator is zero. #/0 is undefined, which is not allowed.

25. $\left(\sqrt{2} - 3\sqrt{6}\right)^2$

   a. $3\sqrt{4}$

   b. $18 - 12\sqrt{3}$

   c. $56 - 12\sqrt{3}$

   d. 50

   e. None of the above

Answer: c

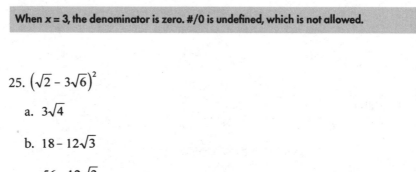

"FOIL" method must be used: First, Outside, Inside, Last.

$$\left(\sqrt{2} - 3\sqrt{6}\right)\left(\sqrt{2} - 3\sqrt{6}\right)$$
$$\sqrt{2} \cdot \sqrt{2} - 3\sqrt{12} - 3\sqrt{12} + 9\sqrt{6} \cdot \sqrt{6} = 2 - 6\sqrt{3} + 54 = 56 - 12\sqrt{3}$$

26. The equation of the ellipse centered at $(1,-3)$ and the tangent to x- and y-axis is:

    a. $(x-1)^2 + (y-3)^2 = 1$

    b. $\dfrac{(x+1)^2}{1} + \dfrac{(y-3)^2}{9} = 1$

    c. $(x-1)^2 + \dfrac{(y+3)^2}{9} = 1$

    d. $(x+1)^2 + \dfrac{(x-3)^2}{3} = 1$

    e. None of the above

Answer: c

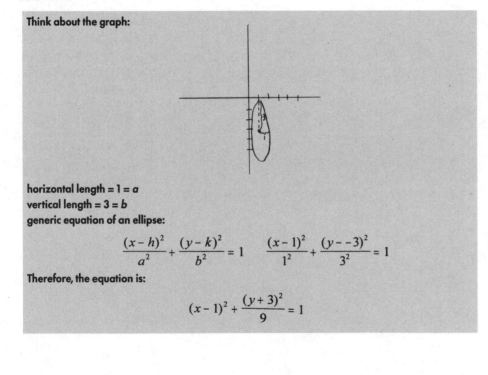

Think about the graph:

horizontal length = 1 = *a*
vertical length = 3 = *b*
generic equation of an ellipse:

$$\frac{(x-h)^2}{a^2} + \frac{(y-k)^2}{b^2} = 1 \qquad \frac{(x-1)^2}{1^2} + \frac{(y--3)^2}{3^2} = 1$$

Therefore, the equation is:

$$(x-1)^2 + \frac{(y+3)^2}{9} = 1$$

27. The graph $f^{-1}(x)$ is found by reflecting the graph of $f(x) = x + 1$ over the:
    a. x-axis
    b. line $y = x$
    c. y-axis
    d. line $y = x + 1$
    e. None of the above

Answer: b

28. Factor $4x^{\frac{1}{2}}$ from $8x^{\frac{5}{2}} + 4x^{\frac{1}{2}}$ leaving

   a. $(2x + 1)$

   b. $2x$

   c. $2x$

   d. $2x + 4$

   e. None of the above

Answer: a

$$\frac{8x^{\frac{5}{2}} + 4x^{\frac{3}{2}}}{4x^{\frac{3}{2}}} = \frac{8x^{\frac{5}{2}}}{4x^{\frac{3}{2}}} + \frac{4x^{\frac{3}{2}}}{4x^{\frac{3}{2}}} = 2x^{\frac{5}{2} - \frac{3}{2}} + 1 = 2x^{\frac{2}{2}} + 1 = 2x + 1$$

CHAPTER 15

INDIANA UNIVERSITY

M014: COLLEGE ALGEBRA

Rasul Shafikov, Associate Instructor

A NUMBER OF QUIZZES, A MIDTERM, AND A FINAL ARE GIVEN IN THIS COURSE. A FINAL exam is presented here.

It's important to understand the theory given in class—basic definitions, statements of main theorems, and so on—but understanding the theory or "getting the idea" is not enough. You should be able to solve the problems, and solve them effectively. You cannot learn math by watching your classmate or an instructor. You have to do the work yourself. Practice: that's the most important ingredient in preparation for the test. Do a lot of sample problems, especially on the topics that you are least comfortable with. It's better to solve one problem, fully understanding the solution, than to solve ten problems by following some standard patterns.

It's certainly good to see some "math culture" in student responses, but taking into account the introductory level of this course, it's a little too much to expect. Correct solutions will do. But if a student also tries to *justify* his or her conclusions, this will make the answer more valuable.

My tests are usually "good student" oriented. Those who are able to complete all the problems in time—and correctly—tend to have excellent skills and a profound understanding of the material. My expectation from an average student is to complete 80 percent of the exam in the time allotted.

All of the problems are focused on different aspects of the course; therefore, all of the questions are equally important. Some of them are easier than others, but that doesn't make them less important.

The key to success in a math course is to work regularly. As a rule, each new topic is based on the previously covered material. Failure to understand one topic may well result in poor understanding of the whole subject. The other important thing is to do a lot of problems. Practice makes perfect.

Try to work out all homework problems yourself. If there is something you cannot solve, it might be a good idea to discuss it with your classmates. If this doesn't clarify the matter, ask your instructor. But avoid asking questions without giving them a thought first. The instructor's answer will probably be forgotten if you don't spend some time trying to figure out the question yourself.

FINAL EXAM

Problem 1. (4%) Evaluate:

$$\left(-\frac{2}{250}\right)^{\frac{1}{3}} =$$

The key idea here is to notice that the fraction $^2/_{250}$ can be reduced to $^1/_{125}$. Now it's easy to finish the problem.

Answer:

$$\left(-\frac{1}{125}\right)^{\frac{1}{3}} = \sqrt[3]{-\frac{1}{125}} = -\frac{1}{5}$$

Problem 2. (4%) Simplify (assume that a, b, and x are positive):

$$\sqrt{25x\sqrt{16x}} =$$

In most problems where radicals are involved, simplifications are easier if we use fractional exponents. Thus, our first and the most important step is to replace square roots with fractional exponents.

$$\sqrt{25x\sqrt{16x}} = \left(25x(16x)^{1/2}\right)^{1/2}$$

Now we use the power rule and the fact that 25 and 16 are perfect squares.

$$\left(25x(16x)^{1/2}\right)^{1/2} = 5\left(x \cdot 4(x)^{1/2}\right)^{1/2} = 10(x^{1+1/2})^{1/2} = 10x^{3/4}$$

Since the original problem was stated using radicals, we can express our final answer in terms of radicals, too.

Answer:

$$10\sqrt[4]{x^3}$$

Problem 3. (6%) Simplify:

$$\frac{x^2 + 2xy + y^2}{x - y} \cdot \frac{y + x}{x^2 - yx} =$$

A common mistake among students is to treat an expression as an equation. While we can add any number to both sides or multiply both sides by a nonzero number in the equation, we cannot do that with an expression without changing its value. For example, in this problem, we cannot multiply the expression by the least common denominator. Instead, we will simplify the expression by canceling equal factors in the numerator and the denominator.

Answer:

$$\frac{x^2 + 2xy + y^2}{x - y} \cdot \frac{y + x}{x^2 - yx} = \frac{(x^2 + 2xy + y^2)x(x - y)}{(x - y)(y + x)} =$$

$$\frac{(x + y)^2 x(x - y)}{(x - y)(y + x)} = x(x + y) = x^2 + xy$$

Problem 4. (6%) Solve the equations:

$$\frac{7x + 1}{3x + 3} - \frac{1}{x + 1} = -2$$

Unlike the previous problem, here we can and should get rid of the denominator by multiplying both sides of the equation by the least common denominator of two fractions on the left-hand side of the equation, i.e., by 3(x + 1).

Answer:

$$\frac{7x + 1}{3x + 3} - \frac{1}{x + 1} = -2$$

$$7x + 1 - 1 \cdot 3 = -2 \cdot 3(x + 1) \Longleftrightarrow$$
$$7x - 2 = -6x - 6 \Longleftrightarrow$$
$$13x = -4 \Longleftrightarrow x = -\tfrac{4}{13}$$

The solution wouldn't be complete without checking that, with this value of x, neither denominator vanishes. It is obvious in this case.

Problem 5. (6%) Evaluate the expression. Your answer should be in the form $ai + b$:

(a) $(5i + 1) - (2i - 4) =$

Answer:

$$(5i + 1) = (2i - 4) = 5i + 1 = 2i + 4 = 3i + 5$$

(b) $\dfrac{2}{1 - i} =$

The division of complex numbers is similar to rationalizing a denominator.

Answer:

$$\frac{2}{1 - i} = \frac{2(1 + i)}{(1 - i)(1 + i)} = \frac{2 + 2i}{1 - i^2} = \frac{2 + 2i}{2} = 1 + i$$

Problem 6. (6%) Solve the inequality. State your answer in interval notation:

$$\frac{x}{x^2 - 1} + \frac{2}{x + 1} \geq 0$$

Solving inequalities is a more delicate matter than solving equations. Denominators are as important as numerators, and we cannot multiply both sides by the least common denominator, since we don't know the sign of this expression. What we need to do is to rewrite the left-hand side as the quotient of two completely factored algebraic expressions:

$$\frac{x}{x^2 - 1} + \frac{2}{x + 1} = \frac{x + 2(x - 1)}{(x - 1)(x + 1)} = \frac{3x - 2}{(x - 1)(x + 1)} \geq 0$$

Answer:

$$x \in \left(-1, \frac{2}{3}\right] \cup (1, \infty)$$

Observe how, in our final answer, we excluded the end points that come from the expressions in the denominator.

Problem 7. (6%)

(a) Show that the line passing through $(1, -1)$ and $(3, 0)$ is parallel to the line $2y - x + 8 = 0$. Your answer should contain at least one full sentence.

First, let's find the slope of the given line. Its equation can be rewritten as follows.

$$y = \frac{1}{2}x - 4$$

Thus the slope is $^1/_2$. Now let's find the slope of the line passing through the two given points.

$$m = \frac{0-(-1)}{3-1} = \frac{1}{2}$$

The important sentence you are expected to say here is the following: *The two lines are parallel because their slopes are equal.* Without this sentence the solution will be incomplete.

(b) Find the equation of a line that passes through the point (3,7) and is perpendicular to the y-axis.

Answer:

The line perpendicular to the y-axis is a horizontal line. The horizontal line passing through the point (3,7) is $y = 7$.

Problem 8. (6%) Find the domain of the function. State your answer in the interval notation.

(a) (5 points) $f(x) = \dfrac{1}{x^2 - 1}$

For the domain we must exclude points for which $f(x)$ is not defined. Usually these are the points where the denominator vanishes or where we have a negative expression under the square root. In this example, the denominator is zero at $x = 1$ and $x = -1$. The answer follows.

Answer:

$$\text{Domain } (f(x)) = (-\infty,-1) \cup (-1,1) \cup (1,\infty).$$

(b) (5 points) $g(x) = \dfrac{1}{\sqrt{x^2 - 1}}$

Here the square root is defined only for $x \in (-\infty,-1] \cup [1,\infty)$. But we need to exclude the end points since the square root is in the denominator. The answer follows.

Answer:

$$\text{Domain } (g(x)) = (-\infty,1) \cup (1,\infty)$$

Problem 9. (4%) If $g(5) = 9$, $(f \circ g)(5) = 21$ find $f(9)$.

Answer:

$$(f \circ g)(5) = f(g(5)) = f(9) = 21, \Rightarrow f(9) = 21$$

Problem 10. (6%) Prove that $f(x) = \sqrt{x} - 1$ and $g(x) = x^2 + 2x + 1$ are inverses of each other.

Answer:

$$f(g(x)) = f(x^2 + 2x + 1) = f((x+1)^2) = \sqrt{(x+1)^2} - 1 = x + 1 - 1 = x$$

Therefore, f and g are inverses of each other.

Problem 11. (6%) Why doesn't the function

$$y = \frac{x^2 - 1}{x^2 + 1}$$

have the inverse function for $x \in (-\infty, +\infty)$?

Justify your answer.

Answer:

This function is not 1–1 by the horizontal line test (or by observing that the function has the same value for $x = 1$ and $x = -1$) and, therefore, it cannot have the inverse function.

This is one of the reasons why one-to-one functions are an important subclass of functions.

Problem 12. (6%) Find a polynomial with integer coefficients of degree 3 with zeros $x = 2i$ and $x = -$. Your answer should be in the form

$$a_n x^n + ... + a_1 x + a_0$$

Since $x = 2i$ is a root, by the Conjugate Roots Theorem so is $x = -2i$. So the polynomial has the following form.

$$(x - 2i)(x + 2i)\left(x + \frac{1}{2}\right) = \left(x^2 + 4\right)\left(x + \frac{1}{2}\right) = x^3 + 4x + \frac{1}{2}x^2 + 2$$

Now we need to multiply the polynomial by an even number, say by 2, to have integer coefficients: $2x^3 + x^2 + 8x + 4$.

Answer:

$$2x^3 + x^2 + 8x + 4$$

This is a possible final answer. In general, different answers are possible here as well. All of them differ by a factor, but it's a good idea to give the simplest one.

Problem 13. (6%) Find:

$$(f \circ g \circ h)(x) \text{ if } f(x) = \frac{1}{x+1}, \ g(x) = \frac{x^2}{2x^2 - 1}, \ h(x) = \sqrt{x}$$

Simplify your answer.

Answer:

$$(f \circ g \circ h)(x) = f(g(\sqrt{x})) = f\left(\frac{\left(\sqrt{x}\right)^2}{2\left(\sqrt{x}\right)^2 - 1}\right) = f\left(\frac{x}{2x-1}\right) = \frac{1}{\frac{x}{2x-1} + 1} = \frac{1}{\frac{x}{2x-1} - \frac{2x-1}{2x-1}} = \frac{2x-1}{-x+1} = \frac{2x-1}{1-x}$$

Note that the order in which we write the composition of two or several functions is very important!

Problem 14. (6%) Find the inverse function of

$$f(x) = \frac{3x^3}{1 + x^3}$$

Answer:

$$y = \frac{3x^3}{1+x^3} \Rightarrow y(1+x^3) = 3x^3 \Rightarrow y + yx^3 = 3x^3 \Rightarrow y = 3x^3 - yx^3 \Rightarrow y = x^3(3-y) \Rightarrow x^3 = \frac{y}{3-y}$$

thus:

$$f^{-1}(x) = \sqrt[3]{\frac{x}{3-x}}$$

Problem 15. (6%) Find the remainder of the following fraction:

$$\frac{6x^{1997} + 4x^2 + 7}{x+1}$$

Answer:

By the Remainder Theorem, the remainder of the preceding fraction is equal to the value

$P(-1)$, where $P(x) = 6x^{1997} + 4x^2 + 7$, $P(-1)^{1997} + 4(-1)^2 + 7 = -6 + 4 + 7 = 5$

Thus, the answer is 5.

You can get the same answer by applying synthetic division, but using the Remainder Theorem is a much faster and nicer way to solve this problem.

Problem 16. (4%) Find all the roots of the polynomial $Q(x) = x3 - 2x^2 - 3x + 6$

Answer:

$$Q(x) = x^2(x - 2) - 3(x - 2) = (x^2 - 3)(x - 2) \Rightarrow Q(x) = \left(x - \sqrt{3}\right)\left(x + \sqrt{3}\right)(x - 2) \Rightarrow$$

The roots of $Q(x)$ are $2, \sqrt{3}$, and $-\sqrt{3}$.

About one-third of the problems on my exams were not discussed in class as stated; therefore, the exams test comprehension of the material rather than some static skills in solving particular types of problems.

Don't spend too much time on a particular exam problem. At the beginning of the test, look through all the problems, find those that seem easy and do them at once. Then try the difficult parts. Try all the problems, even if you don't initially have the faintest idea how to proceed. Guessing is usually the last thing you want to do on a test—but do it if nothing else works!

CHAPTER 16

UNIVERSITY OF NORTH CAROLINA

MATH 10: COLLEGE ALGEBRA

Mark A. McCombs, Director of Teacher Training

THREE MIDTERMS AND A CUMULATIVE FINAL ARE GIVEN IN THIS COURSE. The main course objective is to strengthen the students' algebra and problem-solving skills so they will be better prepared to fulfill the general college math requirement. Required skills include:

◆ Simplifying and combining variable expressions

◆ Solving and graphing equations

◆ Working with functions

◆ Working with mathematical models (one-variable word problems)

The course examinations test these skills with problems that reflect the content and level of difficulty of the assigned text exercise. You should:

◆ Take thorough notes on the main ideas presented in each chapter covered, placing special emphasis on articulating key concepts "in your own words."

◆ Work practice problems and seek help when necessary.

The exams account for 100 percent of the final grade. To do well on them, you should:

◆ Study main ideas in the context of specific problems.

◆ Practice with "closed book, timed" practice tests.

Do all problems!! Show all work!! Label your graphs!!

1. (10 points) Given the equation $y^2 + 3yx = 81$
   a. Find all intercepts (if any).

Answer:

To find the x-intercepts, we set $y = 0$ in the original equation.
$$(0)^2 + 3(0)x = 81$$
$$0 = 81, \text{ which can never happen}$$
So there are no x-intercepts.

To find the y-intercepts, we set $x = 0$ in the original equation:
$$y^2 + 3y(0) = 81$$
$$y^2 = 81$$
$$y = \pm 9$$
So there are two y-intercepts, $(0,9)$ and $(0,-9)$.

b. Determine whether the graph of $y^2 + 3yx = 81$ has origin symmetry. *Show all work!!*

Answer:

To check for origin symmetry, we replace x and y in the original equation by $-x$ and $-y$.
$$(-y)^2 + 3(-y)(-x) = 81$$
Now simplify where possible:
$$y^2 + 3yx = 81, \text{ the negatives all cancel out}$$
And since we get back to the original equation, we have origin symmetry.

2. (10 points)
   a. Draw the graph of line $x = -4$.

Answer:

The given equation represents a vertical line.

$(-4, 0)$

b. Write the equation of the line that is perpendicular to the line $x = -4$ and that passes through the point (6,2).

Answer:

The desired line will be horizontal, passing through (6,2), with equation $y = 2$.

3. (10 points) Find the equation of the line passing through the point (5,–3) and parallel to the line $28x - 4y - 12 = 0$.

We need a point on the desired line as well as the slope of the desired line.

Answer: Point: (5,–3). This is given.

To obtain the desired slope, we need to put the given line in "slope y-intercept form."

Slope: Solve for y in the equation $28x - 4y - 12 = 0$
$$-4y = -28x + 12$$
$$Y = -7x + 3$$
So the given line has slope = 7.
Any line parallel to the given line will also have slope = 7.

Now we can use the "point-slope formula."

$$y - y_1 = m(x - x_1)$$
$$\text{with } (x_1, y_1) = (5, -3) \text{ and } m = 7$$
$$y - (-3) = 7(x - 5)$$
$$y + 3 = 7x - 35$$

final Answer: $y = 7x - 38$

4. (10 points) Find all real solutions for the given equation:
$$6x - 6\sqrt{x} = 5x - 8$$

Step 1: Isolate the radical term.

Answer:
$$6x - 5x + 8 = 6\sqrt{x}$$
$$x + 8 = 6\sqrt{x}$$

Step 2: Square both sides.

$$\left(x + 8\right)^2 = \left(6\sqrt{x}\right)^2$$

$$x^2 + 16x + 64 = 36x$$

Step 3: Solve the resulting quadratic equation.

$$x^2 - 20x + 64 = 0$$
$$(x - 16)(x - 4) = 0$$
$$x = 16 \text{ or } x = 4$$

Step 4: Check answers in the original equation.

Checking $x = 16$:

$$(6)(16) - 6\sqrt{16} = (5)(16) - 8$$
$$96 - 6(4) = 80 - 8$$
$$72 = 72$$

Checking $x = 4$:

$$(6)(4) - 6\sqrt{4} = (5)(4) - 8$$
$$24 - 12 = 20 - 8$$
$$12 = 12$$

Both answers check, so the original equation has solutions $x = 16$ and $x = 4$.

5. (10 points) a. Solve the inequality $x^2 \geq 4(x + 3)$
   Answer on a number line graph.

Answer:

Step 1: Get a zero on one side of the given inequality.

$$x^2 \geq 4x + 12$$
$$x^2 - 4x - 12 \geq 0$$

Step 2: Factor the left-hand side.

$$(x - 6)(x + 2) \geq 0$$

Step 3: Construct a "sign chart," using the x-values that make each factor equal zero.

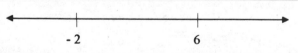

Step 4: Use "test numbers" to determine the sign of the factored expression on each interval.

If $x < -2$, say $x = -3$, then the product $(x - 6)(x + 2)$ becomes $(-)(-)$ = positive.
If $-2 < x < 6$, say $x = 0$, then the product $(x - 6)(x + 2)$ becomes $(-)(+)$ = negative.
If $x > 6$, say $x = 7$, then the product $(x - 6)(x + 2)$ becomes $(+)(+)$ = positive.

So the "sign chart" is

−2   6

Step 5: Examine the original inequality to determine which signs to keep.

In this problem we need the product $(x-6)(x+2)$ to be ≥ 0.
The sign chart indicates the product will be positive for $x < -2$ or $x > 6$.
The product will equal zero for $x = -2$ or $x = 6$.
So the final answer, as a number line graph, is

−2   6

b. (10 points) Solve the inequality
$$\frac{4x+5}{2+x} < 3$$

Answer in interval notation.

Answer:

Step 1: Get a zero on one side of the given inequality.

$$\frac{4x+5}{2+x} - 3 < 0$$

Step 2: Simplify the left-hand side.

$$\frac{4x+5-3(2+x)}{2+x} < 0$$

$$\frac{4x+5-6-3x}{2+x} < 0$$

$$\frac{x-1}{2+x} < 0$$

Step 3: Construct a "sign chart" using the x values that make each factor equal zero.

−2   1

Step 4: Use "test numbers" to determine the sign of the factored expression on each subinterval.

If $x < -2$, say $x = -3$, then the quotient $\frac{x-1}{2+x}$ becomes $(-)/(-) = $ positive.

If $-2 < x < 1$, say $x = 0$, then the quotient $\dfrac{x-1}{2+x}$ becomes $(-)/(+) = $ negative.

If $x > 1$, say $x = 2$, then the quotient $\dfrac{x-1}{2+x}$ becomes $(+)/(+) = $ positive.

So the sign chart:

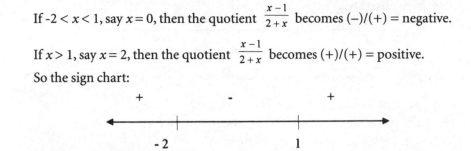

Examine the original inequality to determine which signs to keep.

In this problem, we need the quotient $\dfrac{x-1}{2+x}$ to be <0.

The sign chart indicates the quotient will be negative for $-2 < x < 1$.

The quotient will equal zero for $x = 1$.

The quotient will be undefined for $x = -2$ (since we get $-3/0$).

Final answer, using interval notation: $(-2, 1)$.

6. (10 points) Given the graph:

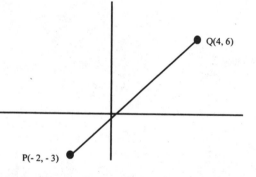

a. Find the slope of the line that is perpendicular to the line containing points P and Q.

Answer: The slope of the given line is

$$\frac{\Delta y}{\Delta x} = \frac{-3-6}{-2-4} = \frac{-9}{-6} = \frac{3}{2}$$

Any line that is perpendicular to this line must have a slope that is the negative reciprocal of 3/2.

Therefore, the $\perp$ *slope* $= -$º.

b. Find the midpoint of segment $\overline{PQ}$.

Answer: Using the midpoint formula, we have:

$$\text{midpoint} = \left(\frac{x_1 + x_2}{2}, \frac{y_1 + y_2}{2}\right) = \left(\frac{-2 + 4}{2}, \frac{-3 + 6}{2}\right) = \left(\frac{2}{2}, \frac{3}{2}\right) = (1, 1.5)$$

7. (10 points) Find the center and radius of the circle with the equation
$$8x^2 + 8y^2 + 160x - 48y - 32 = 0$$

Answer:

Step 1: Complete the square on the given equation.

$$\frac{8x^2 + 8y^2 + 160x - 48y - 32}{8} = \frac{0}{8}$$

$$x^2 + y^2 + 20x - 6y - 4 = 0$$

$$x^2 + 20x + y^2 - 6y = 4$$

$$x^2 + 20x + 100 + y^2 - 6y + 9 = 4 + 109$$

$$(x + 10)^2 + (y - 3)^2 = 113$$

Step 2: Identify the center and radius using $(x - h)^2 + (y - k)^2 = r^2$.

center $(-10, 3)$

radius $= \sqrt{113}$

8. (10 points) Find all real solutions for the equation $x^{2/5} - 7x^{1/5} + 12 = 0$

Answer:

Step 1: Use a substitution to convert the given equation to a quadratic equation.

Let $y = x^{1/5}$; this means that $y^2 = x^{2/5}$, so the original equation becomes $y^2 - 7y + 12 = 0$.

Step 2: Solve the resulting quadratic equation.

We can factor as follows: $(y - 4)(y - 3) = 0$
So we have $y = 4$ or $y = 3$.

Step 3: Express answers in terms of the original variable.

$$y = 4 \text{ means } x^{1/5} = 4$$
$$\text{which means } (x^{1/5})^5 = 4^5$$
$$\text{so } x = 1024$$

$$y = 3 \text{ means } x^{1/5} = 3$$
$$\text{which means } (x^{1/5})^5 = 3^5$$
$$\text{so } x = 243$$

Step 4: Check answers in the original equation.

Checking $x = 1024$:

$$(1024)^{2/5} - 7(1024)^{1/5} + 12 = 0$$

$$\left(\sqrt[5]{1024}\right)^2 - 7\left(\sqrt[5]{1024}\right) + 12 = 0$$

$$(4)^2 - 7(4) + 12 = 0$$

$$16 - 28 + 12 = 0$$
$$0 = 0$$

Checking $x = 243$:

$$(243)^{2/5} - 7(243)^{1/5} + 12 = 0$$
$$\left(\sqrt[5]{243}\right)^2 - 7\left(\sqrt[5]{243}\right) + 12 = 0$$
$$(3)^2 - 7(3) + 12 = 0$$

$$9 - 21 + 12 = 0$$
$$0 = 0$$

Since both answers check, our solutions are $x = 1024$ and $x = 243$.

9a. (6 points) Write the equation of the circle shown here:

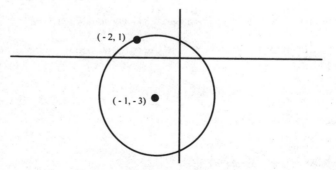

Answer:

Step 1: Use the distance formula to compute the radius.

$$r = \sqrt{(-1-(-2))^2 + (-3-1)^2} = \sqrt{1+16} = \sqrt{17}$$

Step 2: Now use the formula $(x - h)^2 + (y - k)^2 = r^2$.

center point (h,k) = (−1,−3)

radius = $r = \sqrt{17}$

final Answer: $(x + 1)^2 + (y + 3)^2 = 17$

b. (4 points) Draw the graph of the circle with equation $x^2 + y^2 - 9 = 0$

Answer:

Step 1: Rewrite equation in the form $(x - h)^2 + (y - k)^2 = r^2$.

$$x^2 + y^2 = 9$$
$$(x-0)^2 + (y-0)^2 = (3)^2$$

So the center point is (0,0) and the radius = 3.

Step 2: Graph the circle.

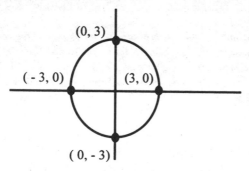

BONUS (5 points)

Since the beginning of the year, the price of gasoline has been decreasing at a constant rate. On January 1st, the price was $1.38 per gallon. On July 1st, the price was $1.19 per gallon. Use $x =$ the number of months elapsed since January 1st to write a linear equation that expresses the relationship between the price of gas and the time of year.

Answer:

Consider the following table:

	x = number of months since Jan. 1st	y = price per gallon
Jan 1st	0	1.38
July 1st	6	1.19

so we have points (0,1.38) and (6,1.19).

We can compute the slope =

$$\frac{\Delta y}{\Delta x} = \frac{1.38 - 1.19}{0 - 6} = \frac{.19}{-6} = -\frac{19}{600}$$

Now use "point slope formula" $y - y_1 = m(x - x_1)$.

$$y - 1.38 = -\frac{19}{600}(x - 0)$$

which means $y = -\frac{19}{600}x + 1.38$

FINAL EXAM

Since each question is equally weighted, I encourage students to spend time first on the problems with which they feel most comfortable. Obviously, the word problems and the complicated graphing problems are the most time-consuming for most students. Students should budget their time according to their awareness of their own mastery of basic concepts and problem-solving techniques relevant to the course.

Instructions:

1. There are 25 problems on this exam. Each problem is worth 8 points.
2. Place your answers in the boxes provided.
3. SHOW ALL WORK ON THESE PAGES. NO CREDIT WILL BE GIVEN FOR CORRECT ANSWERS WITHOUT SUPPORTING WORK.
4. You may use a calculator. However, you MAY NOT use your calculator on the graphing problems on the exam.
5. Read each problem carefully. There will be minimal partial credit given.

1.a. Find all real solutions (if any) for the equation $x - 7 = \dfrac{18}{x}$

Answer:

Step 1: Clear the denominators.

$$(x)(x - 7) = \left(\frac{18}{x}\right)(x)$$

$$x^2 - 7x = 18$$

Step 2: Solve the resulting equation.

$$x^2 - 7x - 18 = 0$$
$$(x - 9)(x + 2) = 0$$
$$x = 9 \text{ or } x = -2$$

Step 3: Check answers in the original equation.

Checking $x = 9$:
$$9 - 7 = 18/9$$
$$2 = 2$$

Checking $x = -2$:
$$-2 - 7 = 18/-2$$
$$-9 = -9$$

So our solutions are $x = 9$ and $x = -2$.

b. Find all real solutions (if any) for the equation: $x(x + 9) = 1$
Give an exact answer and simplify.

Answer:

Step 1: Multiply out the left-hand side.

$$x^2 + 9x = 1$$

Step 2: Solve the resulting quadratic equation.

$$x^2 + 9x - 1 = 0$$

We can use the quadratic formula:

$$x = \frac{-b \pm \sqrt{b^2 - 4ac}}{2a}$$

In this problem, $a = 1$, $b = 9$, and $c = -1$
So we have:

$$x = \frac{-9 \pm \sqrt{9^2 - 4(1)(-1)}}{2(1)} = \frac{-9 \pm \sqrt{85}}{2}$$

2. a. Simplify the radical expression: $\sqrt{162x^{25}y^{12}}$

Answer:

Simplify each term separately, using properties of radicals.

$$\left(\sqrt{162}\right)\left(\sqrt{x^{25}}\right)\left(\sqrt{y^{12}}\right) = \left(\sqrt{(81)(2)}\right)\left(\sqrt{(x^{24})(x)}\right)\left(\sqrt{y^{12}}\right)$$

$$= \left((9)\sqrt{2}\right)\left((x^{12})\sqrt{x}\right)\left(y^{6}\right) = 9x^{12}y^{6}\left(\sqrt{2x}\right)$$

b. Simplify the given expression and write your answer, using only positive exponents:

$$\left(\frac{9x^{7}y^{-10}}{x^{3}}\right)^{3/2}$$

Answer:

Simplify each term separately, using properties of exponents.

$$\left(\frac{9x^{4}}{y^{10}}\right)^{3/2} = \frac{\left(9^{3/2}\right)\left(x^{4}\right)^{3/2}}{\left(y^{10}\right)^{3/2}} = \frac{\left(\sqrt{9}\right)^{3}\left(x^{12/2}\right)}{\left(y^{30/2}\right)} = \frac{27x^{6}}{y^{15}}$$

3. A movie theater sells adult tickets for $7.50 each and children's tickets for $3.50 each. Last Friday, the theater sold a total of 970 tickets. The total deposit from ticket sales was $7031.

How many adult tickets were sold?

Answer:

Let $x =$ the number of adult tickets sold.

Consider the following table:

	Number tickets sold	Price per ticket	Money earned
Adults	x	7.50	7.50x
Children	970 - x	3.50	3.50(970 - x)

Since the total deposit = $7031, we have the equation:

money earned from adult tickets + money earned from children's tickets = 7031

$$7.50x + 3.50(970 - x) = 7031$$
$$7.5x + 3395 - 3.5x = 7031$$
$$4x = 3636$$
$$x = 3636/4 = 909$$

So 909 adult tickets were sold.

4.a. Combine the rational expressions and reduce your answer to the lowest term.

$$\frac{x}{x+6} + \frac{5x}{x-1}$$

Answer in factored form.

Answer:

Step 1: Combine terms using least common denominator.

$$\frac{x(x-1) + 5x(x+6)}{(x+6)(x-1)}$$

Simplify numerator.

$$\frac{x^2 - x + 5x^2 + 30x}{(x+6)(x-1)}$$

$$\frac{6x^2 + 29x}{(x+6)(x-1)}$$

Step 3: Factor and reduce, if possible.

$$\frac{x(6x+29)}{(x+6)(x-1)}$$

b. Write the equation of the line continuing the point P (7,2) and perpendicular to the line $28x + 4y + 8 = 0$.

Answer:

We need a point on the desired line as well as the slope of the desired line.

Point: (7,2). This is given.

To obtain the desired slope, we need to put the given line in "slope y-intercept form." That is, solve for y in the equation $28x + 4y + 8 = 0$.

Slope:

$$4y = -28x - 8$$
$$y = -7x - 2$$

So the given line has slope $= -7$.

Any line perpendicular to the given line will have slope $= \frac{1}{7}$.

Now we can use the "point-slope formula" $y - y_1 = m(x - x_1)$ with $(x_1, y_1) = (7,2)$ and $m = \frac{1}{7}$

$$y - 2 = \frac{1}{7}(x - 7)$$

$$y - 2 = \frac{1}{7}x - 1$$

final Answer:

$$y = \frac{1}{7}x + 1$$

5. Find all real solutions (if any) for the equation $x - \sqrt{x} = 56$

Answer:

Step 1: Isolate the radical term.

$$x - 56 = \sqrt{x}$$

Step 2: Square both sides.

$$\left(x - 56\right)^2 = \left(\sqrt{x}\right)^2$$

$$x^2 - 112x + 3136 = x$$

$$x^2 - 113x + 3136 = 0$$

Step 3: Solve the resulting quadratic equation.

$$(x - 64)(x - 49) = 0$$
$$x = 64 \text{ or } x = 49$$

Step 4: Check answers in the original equation.

Checking $x = 64$

$$64 - \sqrt{64} = 56$$
$$64 - 8 = 56$$
$$56 = 56$$

Checking $x = 49$

$$49 - \sqrt{49} = 56$$
$$49 - 7 = 56$$
$$42 = 56, \text{ not true}$$

Only one answer checks, so the original equation has solution $x = 64$.

6. Given point P on the graph of $y = 16 - x^2$ shown below:

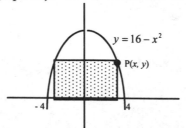

Express the area of the inscribed rectangle as a function of x.

Answer:

Step 1: Express the area in terms of x and y.

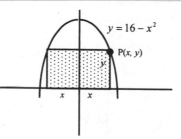

The width of the rectangle $= 2x$ and the height $= y$.
So the area $=$ (length)(width) $= (2x)(y)$.

Step 2: Express all "y values" in terms of x.

Since point P is on the graph of $y = 16 - x^2$, the y-coordinate of point P can be renamed "$16 - x^2$."
So the rectangle area $= (2x)(16 - x^2)$.

7.a. Find all real solutions (if any) for the equation $(3x + 5)(x - 1) = 3x(x + 2)$

Answer:

Step 1: Multiple out each side.

$$3x^2 + 2x - 5 = 3x^2 + 6x$$

Step 2: Isolate the x.

$$-5 = 4x$$

$$-\frac{5}{4} = x$$

b. Write the given inequality using interval notation: $-6 < x \le 4$

Answer: $(-6, 4]$

8. Find all real solutions (if any) for the equation $\dfrac{1}{5-x} + \dfrac{1}{5+x} = 2$
Give exact answer.

Answer:

Step 1: Clear the denominators.

$$(5+x)(5-x)\left(\frac{1}{5-x} + \frac{1}{5+x}\right) = 2(5+x)(5-x)$$

$$(5+x)(1) + (5-x)(1) = 2(25 - x^2)$$

$$5 + x + 5 - x = 50 - 2x^2$$

Step 2: Solve the resulting equation.

$$10 = 50 - 2x^2$$

$$2x^2 = 40$$

$$x^2 = 20$$

$$x = \pm\sqrt{20} = \pm 2\sqrt{5}$$

Step 3: Check answers in original equation.

Since neither answer causes any denominator to equal 0, both answers check.

9. Solve the inequality $\dfrac{2x-6}{1-x} < 1$

Answer:

Step 1: Get a zero on one side of the given inequality.

$$\frac{2x-6}{1-x} - 1 < 0$$

Step 2: Simplify the left-hand side.

$$\frac{2x - 6 - (1 - x)}{1 - x} < 0$$

$$\frac{2x - 6 - 1 + x}{1 - x} < 0$$

$$\frac{3x - 7}{1 - x} < 0$$

Step 3: Construct a "sign chart" using the x-values that make each factor equal zero.

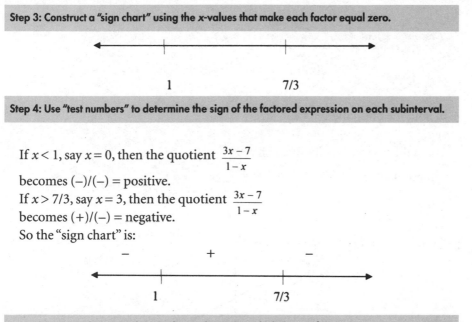

1   7/3

Step 4: Use "test numbers" to determine the sign of the factored expression on each subinterval.

If $x < 1$, say $x = 0$, then the quotient $\frac{3x-7}{1-x}$

becomes $(-)/(-) = $ positive.
If $x > 7/3$, say $x = 3$, then the quotient $\frac{3x-7}{1-x}$
becomes $(+)/(-) = $ negative.
So the "sign chart" is:

       $-$ $+$ $-$

1   7/3

Step 5: Examine the original inequality to determine which signs to keep.

In this problem we need the quotient

$\frac{3x-7}{1-x}$ to be < 0.

The sign chart indicates the quotient will be negative for $x < 1$ or $x > 7/3$.
The quotient will equal zero for $x = 7/3$.
The quotient will be included for $x = 1$, since we get $-7/0$.
Final answer, using interval notation: $(-\infty, 1) \cup (7/3, \infty)$.

10. Give the graph of line segment $\overline{PQ}$

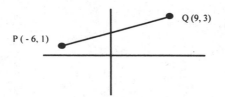

a. Find the distance from point P to point Q. Give an exact answer.

Answer:

Use the distance formula: $d = \sqrt{\left(x_2 - x_1\right)^2 + \left(y_2 - y_1\right)^2}$

$$d = \sqrt{\left(9 - (-6)\right)^2 + (3 - 1)^2} = \sqrt{(15)^2 + (2)^2} = \sqrt{229}$$

b. Find the midpoint segment $\overline{PQ}$.

Answer:

Using the midpoint solution, we have:

$$\text{midpoint} = \left(\frac{x_1 + x_2}{2}, \frac{y_1 + y_2}{2}\right) = \left(\frac{-6+9}{2}, \frac{1+3}{2}\right) = \left(\frac{3}{2}, \frac{4}{2}\right) = (1.5, 2)$$

11. Write the equation of the line shown here:

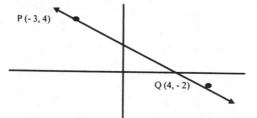

P (- 3, 4)

Q (4, - 2)

Answer in slope-intercept form.

Answer:

We need a point on the desired line as well as the slope of the desired line.

Point: Choose one of the given points, say P(−3,4)

Slope: The slope of the given line is

$$\frac{\Delta y}{\Delta x} = \frac{-2 - 4}{4 - (-3)} = \frac{-6}{7}$$

Now we can use the "point-slope formula" $y - y_1 = m(x - x_1)$ with $(x_1, y_1) = (-3, 4)$ and $m = -6/7$.

$$y - 4 = -\frac{6}{7}(x - (-3))$$

$$y - 4 = -\frac{6}{7}(x + 3)$$

$$y = -\frac{6}{7}x - \frac{18}{7} + 4$$

final Answer: $y = -\frac{6}{7}x + \frac{10}{7}$

12. Graph the piecewise function

$$f(x) = \begin{cases} -5 \text{ for } x < -1 \\ x^3 \text{ for } -1 \le x \le 1 \\ \sqrt{s} \text{ for } 1 < x \le 9 \end{cases}$$

You must label at least four points on your final graph.

Answer:

for $x < -1$, we graph the horizontal line $y = -5$:

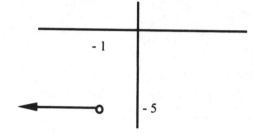

For $-1 \le x \le 1$, we graph the cubic function $y = x^3$:

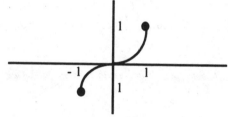

For $1 < x \le 9$, we graph the square root function $y = \sqrt{x}$:

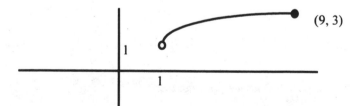

So the final graph is:

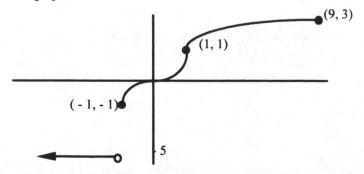

13.a. Find all intercepts (if any) for the equation $\dfrac{12}{x} + 5y - 4 = 0$

Answer:

To find the x-intercepts, we set y = 0 in the original equation.

$$\frac{12}{x} + 5(0) - 4 = 0$$

$$\frac{12}{x} = 4$$

$$12 = 4x$$

$$x = 3$$

So the *x*-intercept is (3,0).

To find the *y*-intercepts, we set $x = 0$ in the original equation.

$$\frac{12}{0} + 5y - 4 = 0$$

So there are no *y*-intercepts, since 12/0 is undefined.

b. Determine whether the equation $\dfrac{12}{x} + 5y - 4 = 0$ has *y*-axis symmetry.

Answer:

To check for y-axis symmetry, we replace x in the original equation by –x.

$$\frac{12}{(-x)} + 5y - 4 = 0$$

Since this is not the same as the original equation, we do *not* have *y*-axis symmetry.

14. Find the center and the radius of the circle $x^2 - 20x + y^2 - 14y - 2 = 0$

Answer:

Step 1: Complete the square on the given equation.

$$x^2 - 20x + 100 + y^2 - 14y + 49 = 2 + 149$$
$$x^2 - 20x + 100 + y^2 - 14y + 49 = 151$$
$$(x - 10)^2 + (y - 7)^2 = 151$$

Step 2: Identify the center and radius using $(x - h)^2 + (y - k)^2 = r^2$
center (10,7)

$$\text{radius} = \sqrt{151}$$

15. Given the function $f(x) = \dfrac{16x + 4}{x^2 + 9x}$

   a. Find the domain for f.

   This function will be undefined whenever its denominator equals 0.

   Solving $x^2 + 9x = 0$, we have
   $$x(x + 9) = 0$$
   $$x = 0 \text{ or } x = -9$$
   so the domain for f is $\{x \text{ such that } x \neq 0,\ x \neq -9\}$.

   b. Find the horizontal asymptote for f (if any):

   Answer:

   To find the horizontal asymptote, we analyze
   $$f(x) = \frac{16x + 4}{x^2 + 9x} \quad \text{as } x \to \pm\infty$$
   That is, we compare the highest power terms in numerator and denominator:
   $$y = \frac{16x + 4}{x^2 + 9x} \approx \frac{16x}{x^2} = \frac{16}{x} \to 0$$
   So the horizontal asymptote is $y = 0$.

16. Given the functions $f(x) = \dfrac{x^2 + 1}{3 - x}$ and $g(x) = \dfrac{10}{x}$

   a. Compute $(f \circ g)(2)$

   Answer:
   $$(f \circ g)(2) = f\big(g(2)\big) = f\left(\frac{10}{2}\right) = f(5) = \frac{(5)^2 + 1}{3 - 5} = \frac{26}{-2} = -13$$

   b. Determine whether $g(x) = \dfrac{10}{x}$ is an odd function.

   Answer:

   An odd function has origin symmetry.

   $$g(x) = \frac{10}{x} \quad \text{means } y = \frac{10}{x}$$

   To check for origin symmetry, we replace x and y in the original equation by $-x$ and $-y$.

   $$(-y) = \frac{10}{(-x)}$$

Now simplify where possible.

$$y = \frac{10}{x}$$

The negatives cancel out.

Since we get back to the original equation, we have origin symmetry. So

$g(x) = \dfrac{10}{x}$ is an odd function.

17. The graph of the function $y = f(x)$ is shown here:

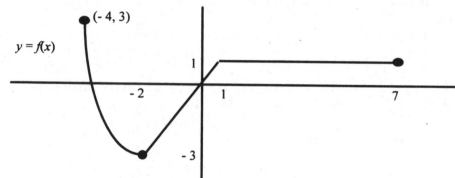

(i) $f(-2) =$
(ii) Is f "one-to-one"?
(iii) domain of f:
(iv) list intervals where f is increasing: $(-2,1)$

Answers:
   (i) $f(-2) = -3$ since the point $(-2,-3)$ is on the graph of $y = f(x)$.
   (ii) No, because the graph of f fails the "Horizontal Line Test."
   (iii) $\{-4 \leq x \leq 7\}$
   (iv) $\{-4 \leq x \leq 7\}$

18.a. An original investment of $4325 earns 7% annual interest compounded quarterly. Find the balance after 9 years. Answer to 2 decimal places.

Answer:

We use the compound interest formula: $A = P\left(1 + \dfrac{r}{n}\right)^{nt}$

with $P = 4325$; $r = .07$; $n = 4$; $t = 9$.

$$A = 4325\left(1 + \frac{.07}{4}\right)^{(4)(9)} = 4325(1.0175)^{36} = 8076.54 \text{ dollars}$$

b. Evaluate the logarithm without using a calculator. The answer must be exact.

$$\log_9\left(\sqrt[7]{81}\right) =$$

Answer:

We can rewrite so that bases match.

$$\log_9\left(\sqrt[7]{\left(9^2\right)}\right) = \log_9\left(9^{2/7}\right) = \frac{2}{7}$$

19.a. Use transformation to draw the graph of the function $f(x) = 4|x + 3|$
   You must label at least three points on your final graph.

Answer:

Step 1: Identify the basic graph.

Here the basic graph is the graph of $y = |x|$:

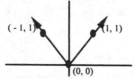

Three points that capture the basic shape are $(-1,1)$, $(0,0)$, $(1,1)$.

Step 2: Apply the transformations, one at a time.

1st transformation: $y = |x + 3|$
We subtract 3 from the x-coordinate of each point on the basic graph: $(-4,1)$, $(-3,0)$, $(-2,1)$.

2nd transformation: $y = 4|x + 3|$
We multiply the y-coordinate of each new point by 4: $(-4,4)$, $(-3,0)$, $(-2,4)$.

Step 3: Plot the final points and draw the basic shape.

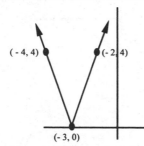

So the original graph was shifted 3 units left and stretched vertically by a factor of 4.

b. Use transformations to draw the graph of the function $f(x) = e^x + 2$
You must label the y-intercept and the asymptote on your final graph.
Answer:

Step 1: Identify the basic graph.

Here the basic graph is of $y = e^x$.

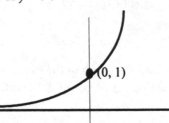

The features that capture the basic shape are:
y-intercept $(0,1)$ horizontal asymptote $y = 0$

Step 2: Apply the transformations, one at a time.

1st transformation: $y = e^x + 2$
We add 2 to the y-coordinate of each feature on the basic graph.
y-intercept $(0,3)$ horizontal asymptote $y = 2$
This is the only transformation.

Step 3: Plot the final points, and draw the basic shape.

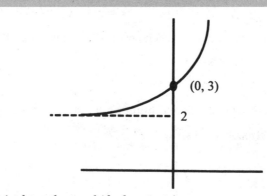

So the original graph was shifted up 2 units.

20.a. Given the graph of the function $y = f(x)$ shown here, draw the graph of the inverse function for f.

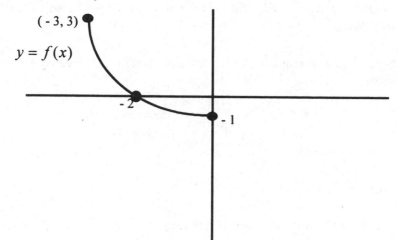

You must label three points on your inverse graph.

To draw the graph of $y = f^{-1}(x)$,
interchange the x- and y-coordinates of the points on the graph of $y = f(x)$.
The graphs of $y = f(x)$ and $y = f^{-1}(x)$ must be symmetric across the line $y = x$.

Answer:

We have the following points:

$(-3,3)$ on the graph of $y = f(x)$ becomes $(3,-3)$ on the graph of $y = f^{-1}(x)$
$(-2,0)$ on the graph of $y = f(x)$ becomes $(0,-2)$ on the graph of $y = f^{-1}(x)$
$(0,-1)$ on the graph of $y = f(x)$ becomes $(-1,0)$ on the graph of $y = f^{-1}(x)$

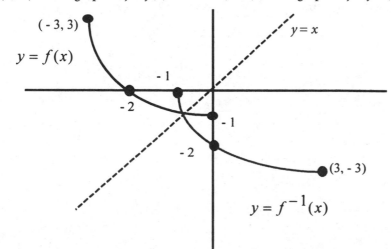

b. The total revenue for a company is given by the function
$R(p) = -987p^2 + 3240p$, where p represents the selling price in dollars.
Find the selling price that will yield the maximum revenue. Answer to 2
decimal places.

Answer:

The revenue function is quadratic, so its maximum value will occur at its vertex.
That is, when

$$p = -\frac{b}{2a} = -\frac{3240}{(2)(-987)} = 1.64 \text{ dollars}$$

21. Given the function $f(x) = \dfrac{9}{x-5}$

a. find the formula for the inverse function for *f.*

Answer:

Step 1: Express the original function as an equation in x and y.

$$y = \frac{9}{x-5}$$

Step 2: Interchange x and y.

$$x = \frac{9}{y-5}$$

Step 3: Solve for y.

Clear the denominator.

$$(y-5)x = \frac{9}{y-5}(y-5)$$

$$yx - 5x = 9$$

isolate the y

$$yx = 9 + 5x$$

$$y = \frac{9+5x}{x}$$

so $f^{-1}(x) = \dfrac{9+5x}{x}$

b. find the range for the original function $f(x) = \dfrac{9}{x-5}$

Answer:

The range of $f(x)$ corresponds with the domain of $f^{-1}(x)$.
The domain of $f^{-1}(x)$ is all x-values not equal to zero.
So the range of $f(x)$ is all y-values not equal to zero.

22. Draw the graph of the quadratic function $f(x) = x^2 + 2x + 5$.
    Identify the vertex and all intercepts.

Answer:

To find the y-intercept, we compute $f(0)$.

$$f(0) = (0)^2 + 2(0) + 5 = 5, \text{ so the } y\text{-intercept is } (0,5).$$

To find the x-intercept(s), we solve $f(x) = 0$.

$$x^2 + 2x + 5 = 0$$

using the quadratic formula, we get:

$$x = \frac{-2 \pm \sqrt{4-20}}{2} = \frac{-2 \pm \sqrt{-16}}{2}$$

which means we have no real solution. So there are no x-intercepts.

To find the vertex, we compute $x = -\dfrac{b}{2a}$

$$x = -\frac{2}{2(1)} = -1$$

So the vertex point $(-1, f(-1)) = (-1,4)$.
The graph is a parabola:

23. a. Find all real solutions to the equation $(5^x)(5^{1-4x}) = 125$
Give an exact answer.

Answer:

Step 1: Rewrite the given equation using matching bases.

In this problem, we can use base 5.
$$(5^x)(5^{1-4x}) = (5)^3$$

Step 2: Simplify, using rules of exponents.

$$(5^{x+1-4x}) = (5)^3$$

Step 3: Set the exponent equal, and solve the resulting equation.

$$x + 1 - 4x = 3$$
$$-3x = 2$$
$$x = -2/3$$

b. Find all real solutions to the equation $\log_3(9x - 2) = 4$
Give an exact answer.

Answer:

Step 1: Rewrite the given equation in exponential form.

$$\log_3(9x - 2) = \text{ becomes } 3^4 = 9x - 2$$

Step 2: Solve the resulting equation.

$$3^4 = 9x - 2$$
$$81 = 9x - 2$$
$$83 = 9x$$
$$83/9 = x$$

Step 3: Check the answer in the original equation.

$$\log_3\left(9\left(\frac{83}{9}\right) - 2\right) = 4$$

$$\log_3(83 - 2) = 4$$

$\log_3(81) = 4$, which is a true statement
So the original equation has solution $x = 83/9$.

24. Draw the graph of the rational function $g(x) = \dfrac{4x^2}{x^2 + 2x - 15}$

Identify all intercepts and asymptotes.

To find the y-intercepts we compute g(0).

$$g(0) = \frac{4(0)^2}{(0)^2 + 2(0) - 15} = 0$$

So the y-intercept is $(0,0)$.

To find the x-intercept(s), we compute g(x) = 0.

$g(x) = \dfrac{4x^2}{x^2 + 2x - 15} = 0$ when the numerator $= 0$

That is, when $x = 0$, so the x-intercept is $(0,0)$.

To find the vertical asymptote(s), we determine when g(x) is undefined.

$g(x) = \dfrac{4x^2}{x^2 + 2x - 15}$ is undefined when the denominator $= 0$.

$$x^2 + 2x - 15 = 0$$

$$(x + 5)(x - 3) = 0$$

So the vertical asymptotes are $x = -5$ and $x = 3$.

To find the horizontal asymptote, we analyze:

$$g(x) = \frac{4x^2}{x^2 + 2x - 15} \quad \text{as } x \to \pm\infty$$

That is, we compare the highest-power term in numerator and denominator.

$$y = \frac{4x^2}{x^2 + 2x - 15} \approx \frac{4x^2}{x^2} = 4$$

So the horizontal asymptote is $y = 4$.

The graph of $y = \dfrac{4x^2}{x^2 + 2x - 15} \approx \dfrac{4x^2}{x^2} = 4$ is:

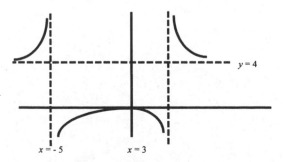

25.a. How long will it take for an investment of $6,100 to grow to a balance of $14,640 if the annual interest rate is 6% compounded continuously? Answer to 2 decimal places.

Continuously compounded interest uses the formula $A = Pe^{rt}$.

Answer:

Given information is:

$$A = 14640; P = 6100; r = .06$$

So we have the equation:

$$14640 = 6100e^{(0.06)t}$$

Now solve for t:

Isolate the exponential term—

$$\frac{14640}{6100} = e^{(0.06)t}$$

$$2.4 = e^{(0.06)t}$$

Apply the natural logarithm—

$$\ln(2.4) = \ln(e^{(0.06)t})$$

$\ln(2.4) = (0.06)t$, since the natural logarithm "undoes" the base e

so we have: $\frac{14640}{6100} = e^{(0.06)t}$ years

b. Find the domain for the function

$$p(x) = \ln(5x - 20)$$

A logarithmic function will be defined provided the expression inside the logarithm is greater than zero.

Answer:

So we solve the inequality $5x - 20 > 0$:

$$5x > 20$$

$$x > 4$$

So the domain for $p(x) = \ln(5x - 20)$ is {all x such that $x > 4$}.

CHAPTER 17

OHIO STATE UNIVERSITY

MATH 104: INTRODUCTORY COLLEGE LEVEL ALGEBRA

Ryan P. Albert, Graduate Teaching Assistant

THREE MIDTERMS AND A FINAL ARE GIVEN IN THIS COURSE. Course objectives include teaching and/or strengthening the algebra skills needed for further math classes and for classes in other areas of study. The student should be able to solve problems of the various types seen in the exams. He or she should be able to factor most common polynomials and understand that algebraically solving problems may lead to solutions that do not make sense in the problem (negative length of sides of a polygon, square root of negative, or dividing by zero).

To succeed in this course, at the very minimum, students must complete all assignments. I would also suggest going through the problem sets in the book and practicing the problems that appear similar to the ones assigned.

Since this course is designed mainly to teach and reinforce problem-solving skills, it is practice and experience that will prove most beneficial.

MIDTERM EXAM

1. Solve for $\dfrac{3f+5}{f+1} = 4$

Answer:

$$3f + 5 = 4(f + 1)$$
$$3f + 5 = 4f + 4$$
$$1 = f$$

2. Solve the inequality $3(x - 5) \leq 4 + 2(2x + 1)$ and give answer in interval notation.
Answer:

$$3x - 15 \leq 4 + 4x + 2$$
$$3x - 15 \leq 4x + 6$$
$$-21 \leq x$$
$$[-21, \infty)$$

3. Solve the inequality $|2x + 3| \geq 5$ and give answer in interval notation.
Answer:

$$2x + 3 \geq 5 \text{ or } 2x + 3 \leq -5$$
$$2x \geq 2 \qquad 2x \leq -8$$
$$x \geq 1 \qquad x \leq -4$$
$$[1, \infty) \cup (-\infty, -4]$$

The "or" is important here. It shows that a solution is possible in one or the other, but not both. No credit given for "and" in this problem.

4. Solve the inequality $-6 \leq 3(x + 2) < 12$ and give answer in interval notation.
Answer:

$$-6 \leq 3 (x + 2) < 12$$
$$-12 \leq 3x < 6$$
$$-4 \leq x < 2$$
$$[-4, 2)$$

The student could have split this into two inequalities using "and," showing that both of them must contain the solution.

5. Solve the system of linear equations:
$$x + 2y = 7$$
$$3x - 4y = 11$$

Either of the following methods could be used. Additional comments are provided in parentheses.

Answer 1:

$$3x + 6y = 21 \text{ (top equation multiplied by 3)}$$
$$3x + 4y = 11$$
$$\Downarrow \text{ (subtracting second from first equation)}$$
$$10y = 10$$
$$y = 1$$
$$\Downarrow \text{ (plugging } y \text{ in to find } x \text{ in first equation)}$$
$$x + 2(1) = 7$$
$$x = 5$$

Answer 2:

$$x = 7 - 2y \text{ (solve for } x \text{ in first equation)}$$
$$3(7 - 2y) - 4y = 11 \text{ (plug in for } x \text{ in second equation)}$$
$$21 - 6y - 4y = 11$$
$$-10y = -10$$
$$y = 1$$
$$\Downarrow \text{ (plugging } y \text{ in to find } x \text{ in first equation)}$$
$$x + 2(1) = 7$$

6. Factor completely:
   I) $2a^4 - 32$
   II) $6b^2 + 7b + 2$

Answer:
   I) $2(a^4 - 16)$
   $2(a^2 - 4)(a^2 + 4)$
   $2(a - 2)(a + 2)(a^2 + 4)$
   II) Factor by grouping:
   Multiply 6 and 2 (12)
   Think of factors of 12 that add to 7 (4,3)
   Rewrite as $6b^2 + 4b + 3b + 2$, then
   $$2b(3b + 2) + (3b + 2)$$
   $$(3b + 2)(2b + 1)$$

7. Reduce to lowest terms:

$$\frac{x^2 + 7x + 12}{x^2 + 3x - 4}$$

Factor both numerator and denominator as follows.

$$\frac{(x+3)(x+4)}{(x+4)(x-1)}$$

Cancel to derive the following.

$$\frac{x+3}{x-1}$$

8. Evaluate and simplify:

$$\frac{6y^4z^3}{5y^2z^5} \div \frac{9y^2z^3}{10y^3z}$$

Answer:

$$\frac{6y^4z^3}{5y^2z^5} \cdot \frac{10y^3z}{9y^2z^3} \;=\; \frac{60y^7z^4}{45y^4z^8} \;=\; \frac{4y^3}{3z^4}$$

9. Evaluate and simplify: $\dfrac{x}{x+2} - \dfrac{3}{x}$

LCD = x(x + 2)

Answer:

$$\frac{x^2}{x(x+2)} - \frac{3(x+2)}{x(x+2)} \;=\; \frac{x^2 - 3(x+2)}{x(x+2)} \;=\; \frac{x^2 - 3x - 6}{x(x+2)}$$

10. Solve the equation: $\dfrac{x}{x-3} + \dfrac{3}{x-2} = \dfrac{-3}{x^2 - 5x + 6}$

LCD = (x – 3)(x – 2). Multiply both sides by LCD to get the following.

Answer:

$$x(x-2) + 3(x-3) = -3 \;\Rightarrow\; x^2 - 2x + 3x - 9 = -3 \;\Rightarrow\; x^2 + x - 6 = 0$$
$$\Rightarrow\; (x-2)(x+3) = 0 \;\Rightarrow\; x = 2, -3$$

But $x = 2$ yields a zero in a denominator in the original problem, so it must be excluded; therefore, $x = -3$.

FINAL EXAM

In every single test answer, I look for the student to prove to me that he or she knows what he or she is talking about. The answer alone will not suffice. Show work and explain what you are doing.

1. Solve for a: $m = \dfrac{ax^2}{3} + rt$

Answer:

$$m - rt = \frac{ax^2}{3} \Rightarrow 3(m - rt) = ax^2 \Rightarrow \frac{3(m - rt)}{x^2} = a$$

2. Find the equation for the line through the point $(1,2)$ and parallel to the line $3x + 5y = 1$ and give answer in slope-intercept form.

Solve for y in the given equation to get the following.

$$y = \frac{-3}{5}x + \frac{1}{5}$$

so that the slope of our line is $\dfrac{-3}{5}$

Using this and point-slope form $y - 2 = \dfrac{-3}{5}(x - 1)$

We solve for y to get the answer that follows.

$$y = \frac{-3}{5}x + \frac{13}{5}$$

3. Solve the following inequality and give the answer in interval notation:
$$-5 < 3x + 8 \le 10$$

Answer:

$$-13 < 3x \le 2 \Rightarrow \frac{-13}{3} < x \le \frac{2}{3}$$

In interval notation, this is: $\left(\dfrac{-13}{3}, \dfrac{2}{3} \right]$

4. Solve the following inequality and give the answer in interval notation: $|3 - 2x| \leq 5$

Answer:

$$3 - 2x \leq 5 \qquad \text{and} \qquad 3 - 2x \geq -5$$
$$-2x \leq 2 \qquad\qquad\qquad -2x \geq -8$$
$$x \geq -1 \qquad\qquad\qquad\quad x \leq 4$$

So the solution set is given by $-1 \leq x \leq 4$, which is $[-1,4]$ in interval notation.

> In this problem, no credit is given for the word "or" used anywhere. This would indicate that the student was not fully aware that the set of solutions *must* be satisfied by *both* of the inequalities rather than by one or the other.

5. Let $f(x) = \dfrac{5 + 4x}{x(x + 8)}$

Specify the domain of $f(x)$ and evaluate $f(-1)$.

Answer:

Domain is all reals, except where the denominator is zero (all real numbers, except 0 and -8).

Plugging -1 in for x, we get:

$$f(-1) = \frac{5 + 4(-1)}{(-1)((-1) + 8)} = \frac{5 - 4}{1 - 8} = \frac{1}{-7}$$

6. Solve the system of linear equations:

$$3x + 2y = 4$$
$$2x - 5y = 9$$

Answer:

$$2(3x + 2y = 4) \qquad \text{yields} \qquad 6x + 4y = 8$$
$$3(2x - 5y = 9) \qquad\qquad\qquad 6x - 15y = 27$$

> Subtract bottom equation from top one to get the following.

$$19y = -19 \Rightarrow y = -1$$

> Now substitute this into one of the original equations and solve for x.

$$x = 2$$

> This problem could have been done as well by the substitution method, as on problem 5 in the midterm exam.

7. Factor completely:

   a) $3a^2 + 14ab - 5b^2$

   b) $3z^2 + 27z^4$

Answer:

   a) $3 \times -5 = -15$ and $14 = 15 - 1$

   so we can rewrite it to be

$$3a^2 + 15ab - ab - 5b^2$$

   and then factor by grouping:

$$3a(a + 5b) - (b(a + 5b) - b(a + 5b)) = (3a - b)(a + 5b)$$

   b) $3z^2(1 + 9z^2)$

> This is the final answer. The question checks to see if the student understands that the difference of squares can be factored, but the sum cannot be.

8. Evaluate and simplify: $\dfrac{3x^3y^5}{10x^2y} \cdot \dfrac{5x^5y^2}{x^7y^6}$

Answer:

   This equals $\dfrac{15x^8y^7}{10x^9y^7} = \dfrac{3}{2x}$

9. Evaluate and simplify: $\dfrac{3}{x^2 + 2x} - \dfrac{4}{x^2 - 3x}$

   LCD is $x(x + 2)(x - 3)$, so this equals

$$\frac{3(x - 3) - 4(x + 2)}{x(x + 2)(x - 3)} = \frac{3x - 9 - 4x - 8}{x(x + 2)(x - 3)} = \frac{-x - 17}{x(x + 2)(x - 3)}$$

10. Solve the equation: $\dfrac{a}{a - 4} - \dfrac{3}{a - 1} = \dfrac{9}{a^2 - 5a + 4}$

Answer:

   LCD is $(a - 4)(a - 1)$

> Multiply both sides by the LCD to get the following.

$$a(a - 1) - 3(a - 4) = 9$$

> Then simplify to get the following result.

$$a^2 - 4a + 3 = 0$$

Now factor.

$$(a-3)(a-1) = 0$$

So $a = 3$ or $a = 1$. But when $a = 1$, there is a zero in one of the denominators; therefore, this answer should be excluded, and $a = 3$ is the only solution.

Eliminating the extra solution is worth about 25 percent of the credit for the problem.

11. Solve the equation: $\sqrt{3x+1} - \sqrt{x+4} = 1$

Answer:

Isolate one radical and square both sides.

$$\sqrt{3x+1} = 1 + \sqrt{x+4} \Rightarrow 3x+1 = 1 + 2\sqrt{x+4} + (x+4) \quad \text{(Must "foil" the RHS.)}$$

$$\Rightarrow 2x-4 = 2\sqrt{x+4} \Rightarrow x-2 = \sqrt{x+4}$$

Now square both sides again to get the following.

$$x^2 - 4x + 4 = x + 4 \Rightarrow x^2 - 5x = 0 \Rightarrow x(x-5) = 0 \Rightarrow x = 0 \text{ or } x = 5$$

But then, in checking these solutions, we find that 0 doesn't work, so $x = 5$. Here checking the solution is worth about 15 percent of the credit for the problem.

$$x = 5$$

12. Evaluate and simplify: $\dfrac{\sqrt[3]{54x^5y^2}}{\sqrt[3]{16x^2y^8}}$

Answer: Rewrite as:

$$\sqrt[3]{\frac{54x^5y^2}{16x^2y^8}} = \sqrt[3]{\frac{27x^3}{8y^6}} = \frac{\sqrt[3]{27x^3}}{\sqrt[3]{8y^6}} = \frac{3x}{2y^2}$$

If this method were not used, then the student would be forced to rationalize the denominator, which would be complicated and messy—and probably lead to some mistakes. If done correctly, however, there would be no problem.

13. Evaluate and simplify: $3\sqrt{48} + 4\sqrt{75} - 10\sqrt{12}$

Answer:

$$3\sqrt{16 \cdot 3} + 4\sqrt{25 \cdot 3} - 10\sqrt{4 \cdot 3} = 3 \cdot 4\sqrt{3} + 4 \cdot 5\sqrt{3} - 10 \cdot 2\sqrt{3} = 12\sqrt{3} + 20\sqrt{3} - 20\sqrt{3}$$

which equals $12\sqrt{3}$ after combining like terms.

14. Rationalize the denominator and simplify: $\dfrac{-3}{\sqrt{7} + \sqrt{2}}$

Answer:

Multiplying by the conjugate of the denominator will get rid of the radicals. The denominator must be "foiled" to get this:

$$\frac{-3}{\sqrt{7} + \sqrt{2}} \cdot \frac{\sqrt{7} - \sqrt{2}}{\sqrt{7} - \sqrt{2}} = \frac{-3(\sqrt{7} - \sqrt{2})}{7 - 2}$$

Which is simplified as much as possible to: $\dfrac{-3(\sqrt{7} - \sqrt{2})}{5}$

15. Let $f(x) = \sqrt{x - 4}$ Find the domain of f.

Answer: $x - 4 > 0 \Rightarrow x \geq 4$

TRIGONOMETRY

CHAPTER 18

INDIANA UNIVERSITY

M026: TRIGONOMETRY

Rasul Shafikov, Associate Instructor

HAPTER 15 PRESENTED A FINAL EXAM IN BASIC ALGEBRA FROM THIS INDIANA University instructor. Here is a midterm exam from another foundation course he teaches—this one in trigonometry.

Problem 1. (10%) Find $\sin \theta$ if $\tan \theta = 4$ and θ is in the first quadrant.

> In the right triangle with sides 1 and 4, the tangent of the angle θ, opposite to 4, equals 4. By the Pythagorean Theorem, the hypotenuse of this triangle will have length $\sqrt{1^2 + 4^2} = \sqrt{17}$. Now it's easy to find the sine of this angle.

Answer:

$$\sin \theta = \frac{opposite}{hypotenuse} = \frac{4}{\sqrt{17}}$$

> Note that this solution is perfectly legitimate, since θ is in the first quadrant.

Problem 2. (10%) In the triangle ABC, $\hat{A} = 9x$, $\hat{B} = 7x$, $\hat{C} = 20°$

Find x. (Here $\hat{C}$ denotes the measure of the angle C.)

> This is a simple application of the Angle Sum of a Triangle Theorem.

Answer:

Since the angle sum of any triangle is 180 degrees, we have the equation:

$$\hat{A} + \hat{B} + \hat{C} = 180° \Leftrightarrow 9x + 7x + 20 = 180$$

Thus, $16x = 160 \Leftrightarrow x = 10°$

The final answer is 10 degrees.

Problem 3. (10%)

a) if tan $\propto$ is undefined, find cot $\propto$.

Answer:

If tan $\propto$ is undefined, then $\alpha = \pm\dfrac{\pi}{2}$. Therefore, cot $\propto$ = 0.

b) if sin β = 0, find cos β.

Answer:

If sin β = 0, then β = ±π, and therefore cos β = ±1.

> Giving just 1 or −1 as the answer is incomplete. Even if the problem doesn't specify that we need to give all possible values of cos, it still implicitly implies that we need to find all such values!

Problem 4. (10%) A traveler left city A and walked 5 miles north. Then he drove 7 miles east; finally, he walked 2 miles south until he reached city B. Find the distance between the cities A and B.

> Plot the route of the traveler on the coordinate plane, assuming that city A is the origin. The coordinates of city B will be (7,3). Now we can easily find the distance between cities A and B using the distance formula.

Answer:

$$dist\,(A, B) = \sqrt{7^2 + 3^2} = \sqrt{58}$$

Problem 5. (10%) Solve the equation $\sin(2x) - \dfrac{1}{\sec(3x-10)} = 0$

Answer:

We have $\sin(2x) = \dfrac{1}{\sec(3x-10)} \Rightarrow \sin(2x) = \cos(3x-10)$.

Now we need to apply the formulas to convert sin to cos.

$$\sin(2x) = \sin(90 - (3x - 10))$$

Now, for simplicity, we can assume that $2x = 90 - 3x + 10 \Leftrightarrow 5x = 100 \Leftrightarrow x = 20$.

The answer, $x = 20$ degrees, is not the best possible answer, but, at this point in the course, it works. The general case will be considered later in the course.

Problem 6. (10%) ABC is a right triangle. $\hat{C} = 90°$, $CA = n$, $CB = m$

Find

a) $\sin A =$

To find sin *A* we need to compute the length of the hypotenuse of *ABC*. This is easily done by Pythagorean Theorem: $AB = \sqrt{n^2 + m^2}$. Now we can answer the question.

Answer:

$$\sin A = \frac{opposite}{hypotenuse} = \frac{m}{\sqrt{m^2 + n^2}}$$

Note that our answer contains only values given in the problem.

b) $\cot B =$

Answer:

$$\cot B = \frac{adjacent}{opposite} = \frac{m}{n}$$

c) $\sec C =$

Answer: $C = 90$ degrees. As we know, cot is undefined for this angle, so the final answer is *undefined*.

Problem 7. (6%) Evaluate without a calculator:

a) (2 pt) sin 270° =

b) (2 pt) sec (−90°) =

Answers:

a) The reference angle to 270 degrees is 90 degrees; sin is negative, thus sin 270° = −1.

b) secant is undefined for −90 degrees.

Problem 8. (5%) Solve the equation: $\sin x = \sqrt{2}$

Answer:

Equation has no solution.

Recall that the range of sine is [−1, 1] and thus $\sqrt{2}$ is not in the range. Therefore this equation has no solution. This is the only possible answer. It is incorrect to say "undefined." We do not define sine here, but ask for a solution to the equation. And the equation either has a solution (unique or not unique) or doesn't.

Problem 9. (5%) Evaluate to the fourth decimal $\dfrac{\tan 7°}{\sin^2 19°} =$

Answer: 1.1498

Use a calculator, and note that you should round your answer to the fourth decimal, using the rule.

Problem 10. (10%) Evaluate without a calculator:

a) sin 120° =

b) cos (−420°) =

Answers:

a) The reference angle to 120 degrees is 60 degrees, and since sin (*x*) is positive in

the second quadrant, we have $\sin 120° = \sin 60° = \dfrac{\sqrt{3}}{2}$.

b) The reference angle is 60 degrees; −420 degrees corresponds to the angle with

the terminal side in the fourth quadrant, where cos (*x*) is positive. Thus,

$$\cos (-420°) = \cos 60° = \frac{1}{2}$$

Problem 11. (6%) Find the area of the square inscribed into a circle of radius R.

Though the problem is stated geometrically, our approach will be analytical, i.e., we will set up an equation to find the area of length of the side of the square.

Answer:

Let a be the side of the square; then, by Pythagorean Theorem,

$(2R)^2 = a^2 + a^2 \Rightarrow 4R^2 = 2a^2$.

Thus $a = \sqrt{2}R$.

The area of a square is then: Area $= a^2 = 2R^2$.

CHAPTER 19

UNIVERSITY OF MISSOURI-ROLLA

MATH 6: TRIGONOMETRY

Stephanie L. Fitch, Lecturer

THREE "MIDTERMS" AND ONE FINAL ARE GIVEN IN THIS COURSE. After successful completion of this course, the student should be prepared to begin the calculus sequence. Over 70 percent of the students at this institution major in engineering, and it is essential for many of them to build up their background in algebra and trigonometry before they are ready to attempt calculus.

This course seems to help students decide if they really want to be engineers or not. I have had a number of students who "want to be engineers" but do not enjoy mathematics, and they begin to realize, as a result of taking this course, that they may be contemplating a career they won't be happy with. Also, nearly all the students in this course are straight out of high school, and many of them are unhappy that they are required to take this course instead of starting the calculus sequence right away. It can be difficult to motivate a student who has this outlook at the beginning of the semester, but, as the course progresses, I think many of the students are glad they spent the time on the basics instead of attempting calculus and having to struggle.

After successful completion of this course, the student will be able to:

◆ Draw and measure angles in standard position

◆ Convert between radian measure and degree measure

◆ Determine the values of trigonometric functions of angles in standard position

◆ Graph the basic trigonometric functions and know how to begin to graph mixed-type functions

◆ Graph and apply inverse trigonometric functions, and know in what intervals they are valid

◆ Verify and use trigonometric identities (students must know the very basic identities: quotient, reciprocal, pythagorean)

◆ Solve trigonometric equations

◆ Apply sum and difference formulas, and multiple angle and product-to-sum formulas (students do not have to memorize the formulas)

◆ Know and be able to apply the laws of sines and cosines, and know when the possibility of more than one answer arises

◆ Translate between the standard form and the trigonometric form of a complex number

◆ Multiply, divide, and take powers and roots of complex numbers (DeMoivre's Theorem)

Also, the student should have developed a sense of how to approach a word problem involving trigonometry, and be able to express an organized thought pattern when solving such problems. Much of the material in the examinations assesses the student's ability to solve the types of problems listed in question 12 in a very direct way. Students are asked to show all of their work, including pictures and diagrams, and to be organized so that their work is easily read and followed. Examinations generally contain at least one application problem in which the student must decide for him- or herself which techniques to use and what formulas may be applicable.

Advice: After each topic is covered in class, work problems from that section until they seem easy. If problems are still a bit difficult, this probably means that you have not yet mastered the material. You may have to do more problems than those that were assigned, but you'll be much more prepared.

Before each test, be sure to work review problems. Being an active studier is the key here. Many students think that if they skim over the book and their old homework, they have studied. In order to retain material, you must approach it actively. If you want to skim over the book, for example, you'll probably find it very helpful to jot down key points, or sketch an outline as you go.

If you begin to get frustrated and are spending what you think is an excessively long time working problems, seek help and advice! Don't wait until you're having serious problems. See your instructor when you have questions. The instructor will be able to give you an idea of about how long your homework should be taking to complete, and give you study tips specifically for his or her class.

MIDTERM EXAM

You have 50 minutes to complete the test. You must *show all work* to receive full credit. If you have any questions, please come to the front and ask.

1. Complete this chart:

Function	Domain	Range
$y = \sin x$		
$y = \sin^{-1} x$		
$y = \cos^{-1} x$		
$y = \tan^{-1} x$		

> Draw graphs of the trig functions. This helps you identify the domain (x-values) and range (y-values). Also, reflecting the graph along the $x = y$ line will produce the graph of the inverse function. Again, you can find the domain and range of the function from its graph.

Answer:

Function	Domain	Range
$y = \sin x$	$\mathfrak{R}$	$-1 \leq y \leq 1$
$y = \sin^{-1} x$	$-1 \leq x \leq 1$	$-\pi/2 \leq y \leq \pi/2$
$y = \cos^{-1} x$	$-1 \leq x \leq 1$	$0 \leq y \leq \pi$
$y = \tan^{-1} x$	$\mathfrak{R}$	$-\pi/2 \leq y \leq \pi/2$

> The answer demonstrates mastery of factual material.
> If the answer is all real numbers, the student must realize the function is valid everywhere. If the answer is an interval, the student must know if the endpoints are included or not, and why.

2. If $\csc x = 5$ and $\cos x < 0$, find the exact values of all six trigonometric functions of x.

> Determine the quadrant in which the cosine is negative. Also notice that the cosecant (hence the sine) must be positive. This will determine the correct quadrant in which to sketch your angle. Label the sides according to the information given, and be sure to include negative signs, if appropriate. Now you can find all the trig functions of your angle right from the drawing.

Answer:

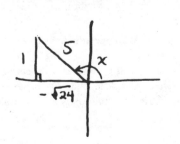

$$\sin x = \frac{1}{5} \qquad \csc x = 5$$

$$\cos x = \frac{\sqrt{24}}{5} \qquad \sec x = \frac{-5}{\sqrt{24}}$$

$$\tan x = \frac{-1}{\sqrt{24}} \qquad \cot x = \sqrt{24}$$

The answer shows the ability to correctly sketch an angle in standard position, and demonstrates knowledge of definition of trig functions.

The student must know in which quadrants the cosine is negative. A drawing is essential. The Pythagorean Theorem must be used to find the length of the sides, and student must know the definitions of the trig functions.

3. Rewrite $\left(\cos^{-1} \dfrac{x}{3} \right)$ as an algebraic expression. Drawing a right triangle will help. Your answer should contain no trigonometric functions.

Since the inverse cosine returns an angle, just substitute an angle θ in its place. You know then that the cosine of θ is $x/3$, so you can sketch a triangle showing θ. Your problem is to find the tangent of θ, which you can determine from your drawing by using the Pythagorean Theorem to label the sides.

Answer:

$$\tan \left(\cos^{-1} \frac{x}{3} \right) = \tan \theta \quad \text{where } \theta = \cos^{-1} \frac{x}{3}, \text{ so } \cos \theta = \frac{x}{3}$$

$$\text{so } \tan \left(\cos^{-1} \frac{x}{3} \right) = \tan \theta = \frac{\sqrt{9 - x^2}}{x}$$

If triangle is drawn as

then $\dfrac{\sqrt{9 - x^2}}{x}$ is acceptable

This response shows knowledge of how to approach an inverse trig function and how to draw a right triangle expressing the problem.

The student must know how to change the problem into something easier to deal with, by realizing that inverse trig functions return an angle that has certain properties. The student must know the possible quadrants in which the inverse cosine may lie. Then a picture is necessary in order to determine the missing side.

4. Verify these identities.

NEVER work on both sides of an identity at once. You should only write an equal sign between expressions you KNOW are equal, not expressions you *want* to be equal.

The best way to approach these types of problems is to copy down the more complicated side first. Then write an equal sign, and write down an expression you KNOW is equal to the original left side. By working in this way, your presentation will follow logically one line to the next. Your last line should be the remaining side of the identity. Sometimes, in these problems, it is helpful to write everything in terms of sine and cosine. Also, if you cannot get to the "other" expression, start over, choosing the opposite side of the equation to work with instead.

a. $\dfrac{\cos x \csc x}{\cot^2 x} = \tan x$

Answer:

$$\frac{\cos x \csc x}{\cot^2 x} = \cos x \cdot \frac{1}{(\sin x)} \cdot \tan^2 x$$

$$= \frac{\cos x}{\sin x} \cdot \frac{\sin^2 x}{\cos^2 x}$$

$$= \frac{\sin x}{\cos x} = \tan x$$

b. $\cos\left(x - \dfrac{\pi}{2}\right) \sec x - \dfrac{1}{\cos x \sec\left(\dfrac{\pi}{2} - x\right)} = 0$

Answer:

$$\cos\left(x - \frac{\pi}{2}\right)\sec x - \frac{1}{\cos x \, \sec\left(\frac{\pi}{2} - x\right)} = \cos\left(-\left(\frac{\pi}{2} - x\right)\right)\frac{1}{\cos x} - \frac{1}{\cos x \csc x}$$

$$= \cos\left(\frac{\pi}{2} - x\right) \cdot \frac{1}{\cos x} - \frac{\sin x}{\cos x}$$

$$= \frac{\sin x}{\cos x} - \frac{\sin x}{\cos x}$$

$$= 0$$

Shows use and knowledge of basic trig identities, and, very importantly, work is shown in an organized manner, in which each line follows directly from the one before.

The student must work through the verification of identities in an organized manner. Every statement written should be a true one, following logically from the previous one. Identities and equal signs must be used correctly.

5. An airplane pilot is following the direction of a highway, at an altitude of 3000 ft. The pilot sees two trucks on the highway ahead. The angle of depression to the farther truck is 20 degrees, and the angle of depression to the closer is 35 degrees. How far apart are the trucks?

> Remember the angle of depression is measured from the top horizontal line downward. Be sure to label your drawing accurately. Also: remember that you can only easily find trig functions of right angles, so be sure you are using the entire side of the large triangle, if that is necessary. At the end, check that you have answered the question that was asked.

Answer:

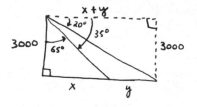

$$\tan 20° = \frac{3000}{x + y}, \text{ so } x + y = \frac{3000}{\tan 20°} \approx 8242.43 \text{ ft}$$

$$\tan 55° = \frac{x}{3000}, \text{ so } x = 3000 \tan 55° \approx 4284.44$$

y = distance between trucks
$= (x + y) - x$
$= 8242.43 - 4284.44$
$= 3957$ ft

> Sketch and labeling show knowledge of how to set up an application problem. Some words are used to explain what the variables stand for. The answer is well organized.
>
> A student who does not draw a picture for this problem is not showing enough work. With an accurate picture that is correctly labeled, the student must then decide how to approach the problem and must go about it in an organized fashion. Also, the student must give the answer that is required, not just the value of a variable that represents the wrong thing.

6. Simplify and match:

  a) $-\cos^2 x$

  b) $\cos^2 x$

  c) $\sec^2 x$

  d) $2\cos x$

  e) $\cos x$

  f) $-\sin x \cos x$

Most of the answers have sine and cosine in them, so change your expression to sines and cosines. Use the basic identities you know to transform the expression into something simpler. Notice that the instructions do not say anything about using an answer more than once.

$$\frac{\csc\ x}{\tan\ x + \cot\ x} =$$

Answer: e

  Work shown:

$$\frac{\dfrac{1}{\sin x}}{\dfrac{\sin\ x}{\cos x} + \dfrac{\cos x}{\sin x}} = \frac{\dfrac{1}{\sin x}}{\dfrac{\sin^2 x + \cos^2 x}{\sin x \cos x}} = \frac{\dfrac{1}{\sin x}}{\dfrac{1}{\sin x \cos x}}$$

$$= \frac{1}{\sin\ x} \cdot \frac{\sin x \cos x}{1} = \cos\ x$$

$$\sin\left(\frac{\pi}{2} - x\right)\cos\ (-x) =$$

Answer: b

  Work shown: $\cos x \cos x = \cos^2 x$

$$\frac{\sin^2 x}{\sec^2 x - 1} =$$

Answer: b

  Work shown:

$$\frac{\sin^2 x}{\tan^2 x} = \frac{\sin^2 x}{\dfrac{\sin^2 x}{\cos^2 x}} = \sin^2 x \cdot \frac{\cos^2 x}{\sin^2 x}$$

$$= \cos^2 x$$

Work is shown to explain the answer, so that partial credit might be possible. The work here shows knowledge of trig identities and algebraic skills.

  The student must be able to use the basic trig identities. Also: the student may be confused, since two answers turn out to be the same. The problem did not state that answers can only be used once, but the student should not be afraid to ask for clarification, if it is needed.

7. Find the *exact* value of cos $(\tan^{-1} 4)$.

Answer:

If $\theta = \tan^{-1} 4$, then cos $(\tan^{-1} 4) = \cos \theta$.

$\tan \theta = 4$

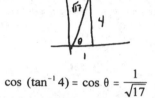

$$\cos\ (\tan^{-1} 4) = \cos \theta = \frac{1}{\sqrt{17}}$$

8. True or false?

a. cos $\theta = \sqrt{1 - \sin^2 \theta}$

Answer: false

b. $\sin^{-1} (\sin x) = x$

Answer: false

c. An object in harmonic motion with displacement $d = 3 \cos \pi x$ has a frequency of 0.5 cycle per unit of time.

Answer: true

d. $3 \sin x = \dfrac{1}{3 \csc x}$

Answer: false

The response shows that the student is not making some common mistakes, and that he or she understands basic definitions.

For part a, the square root is positive while cosine is not always positive. For part b, inverse sin always returns angles in a specific range, whereas x can be any real number. For part c, the student must understand the definition of harmonic motion. For part d, the student must understand that the constant 3 is not really part of the trigonometric expression and should be left alone.

9. Find all solutions of $2\sin^2 x + 3 = 5\sin x$.

Rewrite the equation so that one side is set to zero. Notice it is a quadratic, so you can factor. Now you should have simple expressions involving sin x. Keep in mind that sin x cannot be any number; it is always between −1 and 1. Also be sure you get ALL solutions. (Go around the circle once from your first solution, then twice, etc.)

Answer:

$$2\sin^2 x - 5\sin x + 3 = 0$$
$$(2\sin x - 3)(\sin x - 1) = 0$$
$$\sin x = 3/2 \qquad\qquad \sin x = 1$$
$$\text{can't happen!} \quad x = \pi/2 + 2n\pi$$

Demonstrates recognition of the problem as a quadratic, shows knowledge of range of functions and reasonableness of answers, and gives all possible solutions.

Equation must be set to zero, recognized as a quadratic, factored correctly, and solved for sin x. The student should realize that one solution is extraneous, since it is out of the range of the sine function, and that the other solution gives rise to an infinite number of solutions.

10. Find all solutions of $3t = \sqrt{3}$ in the interval $[0,2\pi)$.

Tangent returns the square root of 3 for what common angle? (There are two!) Draw a picture on the axes of the angles and label the sides. The angle you have drawn is 3t. Then write expressions for ALL solutions for 3t. Next, divide your expressions by 3 so that you have solved for t. Finally, list out the actual values you get from your expressions, and keep only the ones that are in the correct interval. Again, always do this LAST. Write out all the solutions first, when you are still dealing with 3t. Then divide, and, last, select according to your interval. Otherwise, you will lose solutions.

Answer:

$$3t = \begin{cases} \dfrac{\pi}{3} + 2n\pi \\ \dfrac{4\pi}{3} + 2n\pi \end{cases} \xrightarrow{\text{or}} \dfrac{\pi}{3} + n\pi$$

$$\text{so} \quad t = \frac{\pi}{9} + \frac{n\pi}{3}$$

Solution list is $\pi/9, 4\pi/9, 7\pi/9, 10\pi/9, 13\pi/9, 16\pi/9, 19\pi/9, 22\pi/9, \ldots$

In $[0, 2\pi)$, solutions are $\pi/9, 4\pi/9, 7\pi/9, 10\pi/9, 13\pi/9$, and $16\pi/9$

The sketch shows knowledge of how to approach the problem correctly, and division by 3 is not done until AFTER all solutions are listed. Narrowing of solution set is not done until the end, which is correct.

The student should draw a picture to show knowledge of where possible solutions may lie, and then solve for 3*t*, getting all solutions. Next, the student should divide to solve for *t* alone, and, last, list out the solutions that are in the specified range.

FINAL EXAM

I have found that sample tests are a great help to some students. Some instructors provide such tests, but students are perfectly capable of making up sample tests on their own. It is best for a student to select questions out of a book, since the answers are more predictable. When the student "takes" the sample test, he or she should be sure to be in a good test-taking environment, in a quiet area, and under a time constraint. When the test is completed, the student can take it to the instructor for grading. Most instructors will be impressed by the student's resourcefulness and will agree to grade the sample test.

I have found this to be a great way for students to gauge mastery of the material. They have a good idea of how they will perform, and what material they need to practice. Also, this seems to have either a calming effect or a wake-up effect on the student, both of which are valuable.

1. Complete the table. Use exact values.

θ	$\sin \theta$	$\cos \theta$	$\tan \theta$	$\cot \theta$	$\sec \theta$	$\csc \theta$
$\dfrac{-\pi}{3}$						
	$\dfrac{1}{(\leq \theta \leq 2\pi)}$					
$150°$						
$\dfrac{\pi}{4}$						

Know the trig function values for the common angles, or at least be able to derive them by drawing triangles.

Answer:

θ	$\sin\theta$	$\cos\theta$	$\tan\theta$	$\cot\theta$	$\sec\theta$	$\csc\theta$
$\dfrac{-\pi}{3}$	$\sqrt{3}/2$	$1/2$	$\sqrt{3}$	$-1/\sqrt{3}$	2	$-2/\sqrt{3}$
$\pi/2$ or $90°$ ($\leq\theta\leq2\pi$)	$\dfrac{1}{}$	0	Undefined	0	Undefined	1
$150°$	$1/2$	$\sqrt{3}/2$	$-1/\sqrt{3}$	$\sqrt{3}$	$-2/\sqrt{3}$	2
$\dfrac{\pi}{4}$	$1/\sqrt{2}$	$1/\sqrt{2}$	1	1	$\sqrt{2}$	$\sqrt{2}$

The answer demonstrates command of factual material. Student must show knowledge of when trig functions are negative and positive, and have either memorized or be able to find the values for basic angles.

2. Find 3 values of θ if

   a. $\sec\theta=\dfrac{-2}{\sqrt{3}}$

Answer:

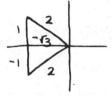

$\theta=150°,210°,510°,\ldots$
or
$\theta=5\pi/6,\ 7\pi/6,\ 17\pi/6,\ldots$

   b. $\cot\theta=1$

Answer:

$\theta=45°,225°,405°,\ldots$
or
$\theta=\pi/4,\ 5\pi/4,\ 9\pi/4,\ldots$

For part a, recall in which quadrants secant (think cosine) is negative. There are two, so draw two possible triangles and label them. These triangles should be familiar, so you can get the basic angles from here. Since you are asked for three angles, think of going around the circle again. A similar process should be used for part b.

   The answer here demonstrates ability to translate the equation into a drawing correctly. Note that the student correctly shows more than one possible triangle. The student must know which quadrants are appropriate, must draw a picture to show this, and must give the required number of solutions.

3. Evaluate the following exactly.

Remember that inverse trig functions return angles in a particular range. Drawing triangles in the correct quadrants and labeling them may help. Also, for part c, remember that inverse sine and sine do not necessarily cancel each other out.

   a. $\arcsin\left(-\dfrac{1}{\sqrt{2}}\right) =$

Answer:

$$45° \text{ or } {}^{-\pi}\!/_4$$

   b. $\csc\left(\tan^{-1}\left(\dfrac{-5}{12}\right)\right) =$

Answer:

$${}^{-13}\!/_5$$

   c. $\sin^{-1}(\sin(3\pi)) =$

Answer: $\sin^{-1} 0 = 0$

Correct factual material is furnished. Also, part b exhibits knowledge of how inverse trig functions can be translated into a drawing.

   Students must understand domain and range of the inverse trig functions, realize that values of these are angles, and be able to draw sketches of these angles, if necessary, to show understanding. Students must also realize in part c that inverse sine does not always "cancel out" sine.

4. Use the given values and trigonometric identities to find the indicated trigonometric functions given $\cos \alpha = 1/4$, α in the 4th quadrant.

> Sketch the appropriate diagram. Draw α in the fourth quadrant and use the information given to label the sides of your right triangle. Be sure to use negative values if you need to. This will give you the answers to the first three parts. For part d, recall the cofunction identities.

  a. $\sec \alpha =$
Answer: 4

  b. $\sin \alpha =$
Answer: $\dfrac{\sqrt{15}}{4}$

  c. $\cot \alpha =$
Answer: $\dfrac{-1}{\sqrt{15}}$

  d. $\sin(90° - \alpha)$
Answer: $\cos \alpha = 1/4$

> The response demonstrates knowledge of basic definition of cosine in order to sketch. The answer shows how to interpret the sketch using basic definitions, and shows knowledge of a basic trig identity.
>   The student must draw a sketch of the angle in the correct quadrant and label it correctly using the Pythagorean Theorem. Then he or she must use knowledge of basic definitions to determine solutions. For part d, the student must know the cofunction identity.

5. Given that $\sin x = \dfrac{\sqrt{33}}{7}$ and $\cos x = \dfrac{4}{7}$, find the exact value of the other four
   functions and the exact value of $\tan 2x$.

> You know the sine is negative and the cosine is positive, which will help you determine the correct quadrant in which to draw the angle x. Label your drawing, using negatives where necessary, and use the drawing to find values of trig functions. For the last part, use the (given) identity and plug in the value of tangent of x to achieve the answer.

Answer:

$$\tan x = \sqrt{33}\big/4$$

$$\csc x = {}^{-7}\big/\sqrt{33}$$

$$\sec x = 7\big/4$$

$$\cot x = {}^{-4}\big/\sqrt{33}$$

$$\tan 2x = \frac{2\tan x}{1 - \tan^2 x}$$

$$= \frac{\sqrt{33}\big/2}{1 - {}^{33}\big/16}$$

$$\frac{\sqrt{33}\big/2}{-17\big/16\ 8} = \frac{8\sqrt{33}}{17}$$

Shows knowledge of how to position an angle when signs of trig functions are known. Also demonstrates knowledge of definitions of trig functions, and shows ability to apply a formula in a particular case. (Formulas are attached to the back of the test and are accessible to students.)

   It is necessary for students to use the negative/positive values of sine and cosine to determine the correct quadrant of the angle, and then draw a correctly labeled sketch so they can use the information to get values for the other trig functions. For the last part, the student must use the (given) identity and fill in values correctly, and simplify the answer.

6. Sketch the graph of $y = 4\cos 2\left(x + \dfrac{\pi}{4}\right)$ Include two full periods.

One easy way to approach this type of problem is to realize that you can simply sketch the general shape of the cosine graph and then label the important points on the graph. We generally begin a drawing of cosine at $x = 0$. Decide what value of x will give an angle that is zero for this problem, and begin your graph there. Similarly, at what x-value is one cycle of the cosine graph usually completed? What value of x will give that angle for this problem? The amplitude is obviously 4, so now you are ready to draw your sketch. Remember to draw as many cycles as the problem requires.

Answer:

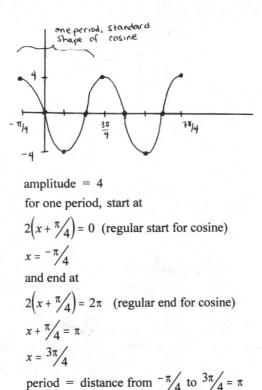

amplitude $= 4$

for one period, start at

$2\left(x + \dfrac{\pi}{4}\right) = 0$ (regular start for cosine)

$x = \dfrac{-\pi}{4}$

and end at

$2\left(x + \dfrac{\pi}{4}\right) = 2\pi$ (regular end for cosine)

$x + \dfrac{\pi}{4} = \pi$

$x = \dfrac{3\pi}{4}$

period $=$ distance from $\dfrac{-\pi}{4}$ to $\dfrac{3\pi}{4} = \pi$

The response shows the correct shape of a cosine graph, knowledge of amplitude, and an organized method of determining pertinent points on the graph. The graph must be neat and clear, and must show at least two periods. It must be labeled correctly, showing amplitude and important x-values. Also, the student should show some kind of work to demonstrate how these correct "starting" and "ending" places were found.

7. Find all solutions of the following equation: $2\sin(3x) + \sqrt{3} = 0$

> Begin by manipulating the equation to get the trig function alone on one side. The value for the sine is one you should recognize. Draw a sketch of the appropriate triangles on the coordinate axes, being sure to label the sides. Now you can solve the problem for the angle given, namely $3x$. Write the solution for $3x$ so that all possible angles are shown. Now simply divide by 3 to find x. It is very important here to list all the solutions before dividing by 3, since you will leave out two-thirds of the solutions if you only add on the $2n\pi$ term at the very end.

Answer:

$$\sin 3x = \frac{\sqrt{3}}{2}$$

$$3x = \begin{cases} 4\pi/3 + 2n\pi \\ 5\pi/3 + 2n\pi \end{cases}$$

$$x = \begin{cases} 4\pi/9 + 2n\pi/3 \\ 5\pi/9 + 2n\pi/3 \end{cases}$$

> The answer shows that the student can sketch angles corresponding to the problem, and that he or she also knows there will be two of them. The response shows division by 3 last, after adding on the extra period to get all solutions—so it is clear that the student understands this important concept.
>
> First solve for $\sin(3x)$, and then sketch both possible angles $3x$. These must be recognized as common angles, and all possible solutions for $3x$ must be given. Last, divide by 3 to solve for x.

8. Solve on $[0,2\pi)$. $2\sin^2 2x + 5\sin 2x - 3 = 0$

> Recognize this as a quadratic equation, and factor. Then solve for the trig function. Remember what values can be returned for the sine, so you can determine whether answers are extraneous. Now, as in the previous problem, draw the correct triangles and label them appropriately so you can identify the angles for 2x. Again, list out all solutions, then divide by 2 to get answers for x. When the answers are to be restricted as in this problem, solve first without the restriction and then list out your solutions. As the final step, only include solutions in the appropriate range.

Answer:

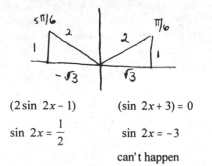

$$(2\sin 2x - 1) \qquad (\sin 2x + 3) = 0$$

$$\sin 2x = \frac{1}{2} \qquad\qquad \sin 2x = -3$$

$$\text{can't happen}$$

$$2x = \begin{cases} \pi/6 + 2n\pi \\ 5\pi/6 + 2n\pi \end{cases}$$

$$x = \begin{cases} \pi/12 + n\pi \\ 5\pi/12 + n\pi \end{cases} \xrightarrow[\text{so, in } [0,2\pi)]{} x = \begin{cases} \pi/12, \; 13\pi/12,\dots \\ 5\pi/12, \; 17\pi/12,\dots \end{cases}$$

> Demonstrates recognition of the problem as a quadratic, as well as knowledge of range of sine and knowledge of how to draw angles corresponding to the problem. Also, all solutions are shown, then division by 2, then culling of solutions to the ones in the specified range—all as it should be.
>   See comments on midterm question 9.

9. Find the exact value of the following:
    a. $\sin 195°$

Answer:

$$= \sin(150° + 45°) = \sin 150° \cos 45° + \cos 150° \sin 45°$$

$$= \left(\frac{1}{2}\right)\left(\frac{1}{\sqrt{2}}\right) + \left(\frac{-\sqrt{3}}{2}\right)\left(\frac{1}{\sqrt{2}}\right)$$

$$= \frac{1 - \sqrt{3}}{2\sqrt{2}}$$

b. cos 105°

Answer:

$$= \cos\ (45° + 60°) = \cos\ 45° \cos\ 60° - \sin\ 45° \sin\ 60°$$

$$= \left(\frac{1}{\sqrt{2}}\right)\left(\frac{1}{2}\right) - \left(\frac{1}{\sqrt{2}}\right)\left(\frac{\sqrt{3}}{2}\right) = \frac{1 - \sqrt{3}}{2\sqrt{2}}$$

> Recognize the angles given as ones that are not necessarily common angles. Then change them into a sum or product of common angles and use an appropriate formula.
>
> The response shows the correct application of a formula to a specific problem, and shows knowledge of the important angles to look for.
>
> The student should recognize the angles as non-common, and realize they must be dealt with as sums, differences, or multiples of common angles. Then simple application of the formulas (which are given) is required.

10. Find the values of the other five trigonometric functions of θ if csc θ = 5, cot θ < 0.

> First, identify the correct quadrant of θ. Where is cosecant positive (same as asking where is sine positive), and where is cotangent negative? Sketch the appropriate triangle on the coordinate axes and correctly label, remembering to use negative numbers if necessary. From the drawing, use the basic definitions of the trig functions to determine the values asked for.

Answer:

$$\sin\ \theta = \frac{1}{5}$$

$$\cos\ \theta = \sqrt{24}\Big/5$$

$$\tan\ \theta = 1\Big/\sqrt{24}$$

$$\sec\ \theta = {-5}\Big/\sqrt{24}$$

$$\cot\ \theta = \sqrt{24}$$

> Shows correct placement of angle on sketch, with correct labeling, and knowledge of definition of trig functions. See comment on midterm question 2.

11. Find the exact value of $\sin 59°\cos 14° - \cos 59°\sin 14°$.

Notice that the angles in this problem are not common ones, but their difference is. Use the appropriate formula.

Answer:

$$= \sin\ (59° - 14°)$$

$$= \sin\ 45°$$

$$= \frac{1}{\sqrt{2}}$$

Shows correct application of formula to a specific problem, and shows knowledge of important angles to look for. Compare final exam question 9.

12. Verify:

a. $\dfrac{1 + \sin\ x}{\cos\ x \sin\ x} = \sec\ x(\csc\ x + 1)$

Answer:

$$\frac{1 + \sin\ x}{\cos\ x \sin\ x} = \left(\frac{1}{\cos\ x}\right)\left(\frac{1 + \sin\ x}{\sin\ x}\right)$$

$$= \sec\ x\left(\frac{1}{\sin\ x} + \frac{\sin\ x}{\sin\ x}\right)$$

$$= \sec\ x(\csc\ x + 1)$$

b. $\dfrac{1}{\cos\ x \sin\ x} + \dfrac{1}{\tan\ x + 1} = 1$

Answer:

$$\frac{1}{\cot\ x + 1} + \frac{1}{\tan\ x + 1} = \frac{\tan\ x + 1}{(\cot\ x + 1)(\tan\ x + 1)} + \frac{\cot\ x + 1}{(\cot\ x + 1)(\tan x + 1)}$$

$$= \frac{\tan\ x + \cot\ x + 2}{\cot\ x \tan\ x + \cot\ x \tan\ x + 1} = \frac{\tan\ x + \cot\ x + 2}{1 + \cot\ x + \tan\ x + 1}$$

$$= \frac{\tan\ x + \cot\ x + 2}{\tan\ x + \cot\ x + 2} = 1$$

c. $\sec\ x = \dfrac{\sec^2 x \cot\ x - \cot\ x}{\sin\ x}$

Answer:

$$\frac{\sec^2 x \cot x - \cot x}{\sin x} = \frac{\cot x(\sec^2 x - 1)}{\sin x}$$

$$= \frac{\cot x(\tan^2 x)}{\sin x} = \frac{\tan x}{\sin x}$$

$$= \frac{\sin x}{\tan x} \cdot \frac{1}{\sin x}$$

$$= \frac{1}{\cos x} = \sec x$$

The response exhibits knowledge of trig identities and also shows algebra skills. More importantly, good organization is shown, and each line follows from the previous one. See midterm question 4.

13. Simplify to a single function of a single angle: $\dfrac{\sin (2x)}{2 \tan x} - \cos (2x)$

First, decide if you want to change the entire problem into an expression in x or in $2x$. Which seems more reasonable, using the formulas to which you have access? Since you are to have only one trig function in your answer, it is likely to be helpful to change everything into sines and cosines.

Answer:

$$\frac{\sin (2x)}{2 \tan x} - \cos (2x) = \frac{2\sin x \cos x}{2 \tan x} - (\cos^2 x - \sin^2 x)$$

$$= (\sin x \cos x)\left(\frac{\cos x}{\sin x}\right) - \cos^2 x + \sin^2 x$$

$$= \sin^2 x \qquad \text{or} \qquad 1 - \cos^2 x$$

Demonstrates knowledge of how to apply formulas and work toward a goal. Also shows knowledge of basic trig identities and algebra skills.

It is necessary to choose whether to change all angles to x or change them all to $2x$. The student should realize it is easier to change to x. Then correct application of identities is required.

14. Write the complex number $-2\sqrt{3} - 2i$ in trigonometric form.

Answer:

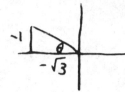

$$r = |-2i - 2\sqrt{3}| = \sqrt{(-2)^2 + (-2\sqrt{3})^2} = \sqrt{4 + 12} = 4$$

$$\tan \theta = \frac{b}{a} = \frac{-2}{-2\sqrt{3}} = \frac{1}{\sqrt{3}}$$

$z = -2\sqrt{3} - 2i$ is in Quadrant III, so $\theta = \dfrac{5\pi}{6}$

$$z = r(\cos \theta + i \sin \theta) = 4(\cos \frac{5\pi}{6} + i \sin \, {}^{5\pi}\!/_{6})$$

15. Given the complex numbers
$$z_1 = 3\left(\cos \frac{4\pi}{3} + i \sin \frac{4\pi}{3} \right), \quad z_2 = 4\left(\cos \frac{\pi}{4} + i \sin \frac{\pi}{4} \right),$$
find the following. Leave your answer in trigonometric form.

a. $z_1 z_2$

Answer:

$$= 12\left(\cos \frac{19\pi}{12} + i \sin \frac{19\pi}{12} \right)$$

$$\uparrow \qquad \uparrow \qquad \qquad \uparrow$$

$$r_1 \cdot r_2 \quad \theta_1 + \theta_2$$

b. z_1 / z_2

Answer:

$$= \frac{3}{4}\left(\cos \frac{13\pi}{12} + i \sin \frac{13\pi}{12}\right)$$

$$\uparrow \qquad \uparrow \qquad\qquad \uparrow$$

$$\frac{r_1}{r_2} \qquad \theta_1 - \theta_2$$

c. $(z_1)^4$

Answer:

$$= 3^4\left(\cos 4\left(\frac{4\pi}{3}\right) + i \sin 4\left(\frac{4\pi}{3}\right)\right)$$

$$= 81\left(\cos \frac{16\pi}{3} + i \sin \frac{16\pi}{3}\right)$$

d. The square roots of z_2

Answer: Roots (there are two):

$$\sqrt{4}\left(\cos \frac{\pi/4}{2} + i \sin \frac{\pi/4}{2}\right) = 2\left(\cos \frac{\pi}{8} + i \sin \frac{\pi}{8}\right)$$

and

$$\sqrt{4}\left(\cos \frac{\pi/4 + 2\pi}{2} + i \sin \frac{\pi/4 + 2\pi}{2}\right) = 2\left(\cos \frac{9\pi}{8} + i \sin \frac{9\pi}{8}\right)$$

This question simply asks the student to recall appropriate formulas for dealing with complex numbers in trigonometric form. The response shows knowledge of formulas for complex numbers and how to apply them. The formulas must be applied correctly, and, for part d, the student must realize there are two roots.

16. A triangle is to be made out of three sticks of wood that are 5, 8, and 9 inches long, respectively. What angle should the first two sticks (5" and 8") make in order that the third one will be just long enough to join their ends?

First, make a sketch of the triangle, and be fairly accurate. Sometimes you can tell from a drawing that the triangle you are trying to draw does not exist! This one does, however, and because of the type of information given (three sides), you should use the law of cosines to determine the angle required.

Answer:

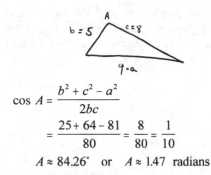

$$\cos A = \frac{b^2 + c^2 - a^2}{2bc}$$

$$= \frac{25 + 64 - 81}{80} = \frac{8}{80} = \frac{1}{10}$$

$$A \approx 84.26° \quad \text{or} \quad A \approx 1.47 \text{ radians}$$

The sketch shows understanding of the problem and is drawn accurately enough not to mislead the student. The response shows knowledge of the law of cosines and when it should be applied.
   The drawing should be accurate enough so that the triangle looks possible. The student should know how to use law of cosines and recall the correct formula.

17. As you travel across flat land, you see a mountain directly in front of you. The angle of elevation to the peak is 4 degrees. After you travel 13 miles toward the mountain, the angle of elevation is 10 degrees. Approximate the height of the mountain.

A sketch is essential. Remember the definition of an angle of elevation. Be sure to label your drawing as is appropriate. You can try to relate any two sides you like with trig functions, but it makes the most sense to use tangent since you are given some information about the horizontal and want to find the vertical distance. When you arrive at an answer, be sure to read the problem again and be sure you have given the information required.

Answer:

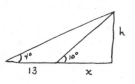

$$\tan 4° = \frac{h}{13 + x,} \qquad h = (\tan 4°)(13 + x)$$

$$\tan 10° = \frac{h}{x,} \qquad h = (\tan 10°)(x)$$

$$x \tan 10° = (13 + x) \tan 4°$$

$$x \tan 10° - x \tan 4° = 13 \tan 4°$$

$$x = \frac{13 \tan 4°}{\tan 10° - \tan 4°} \approx 8.54 \text{ miles}$$

A drawing is necessary, demonstrating knowledge of what *angle of elevation* means. The drawing must be labeled accurately. The answer should be the height, not some other value labeled on the drawing. Sketch shows good understanding of the problem and efficient labeling, the work is organized, and units are given in the answer, as required.

CALCULUS

CHAPTER 20

MA 123: CALCULUS I

Emma Previato, Professor

TRADITIONALLY, ALL INSTRUCTORS IN THIS COURSE GIVE THREE OR FOUR ONE-HOUR exams during the semester. I try to give four, so as to be able to drop the lowest score and allow for absences. The larger number also has the advantage that the students review and are tested often, splitting the material into four units as opposed to three. A final exam is also given. An example of one of the hour exams and the final are included here.

This is an introductory calculus course, and the majority of the students are science or engineering, as opposed to mathematics, majors. Thus, I focus on the applications of calculus, as opposed to the theory. The main goal is for the students to be able to do the problems from the textbook; ideally, they should be able to set them up for themselves, as if faced with a real-life situation. This is why I insist on the (more challenging) "word problems." As for the theoretical understanding, I insist on the concept of a function, its graph, and the operations of calculus (differentiation and integration) as relating to its geometry.

The problems I put on the exams are similar to those found in the course textbook. I want students to walk into the exam confident that, if they have worked and learned how to solve those problems, they can perform well.

It is important to:

1. Do the homework each day as soon as possible after class. Before reading the theory, read the questions for focus.

2. Talk with other students first about anything that is unclear; mathematics gains immensely from being verbalized, and the value of collaboration cannot

be overemphasized. Then, if necessary, ask a tutor (a free service provided at our university, staffed by a graduate or undergraduate student) or the instructor for the solutions. I post the solutions in the library as well.

I give a short list of review problems prior to each exam (about a week before, so that the students can still ask questions); each question on the exam is exemplified by a set of review problems (taken from the homework). Then, I recommend that a student give himself or herself a "practice exam," based on those review problems, especially timing the performance. The most successful class I ever held was one to which I was able to give such a practice test myself (attendance was totally optional). Unfortunately, there is not enough time for that to be common practice.

The following is from the course syllabus:
The math skills you learn in this course form the basis of your scientific competence. Here is what you want to get out of the course: (1) learn how to put a real-life ("word") question into a mathematical model; (2) learn how to manipulate the math to get an answer; (3) *understand and retain* the basic principles used in the previous steps (such as graphing and interpreting a function)— this is why I don't allow crib sheets during exams. The way to achieve these goals is to solve as many problems as possible. In class, I explain the basic principles for each new topic; then I go over samples of the problems I assign as homework. I assign a small number of problems, because the quizzes test you only on assigned problems! However, this forces me to skip the easiest problems in each section; keep in mind that some assigned problems are multiple-step; for practice or with a tutor, do some of the lower-numbered questions. **DO THE HOMEWORK EACH TIME,** and you'll achieve excellence. If you can't do a problem, ask for help! Bear in mind that, during exams and quizzes, you will have to work entirely by yourself (books closed); so, if someone shows you how to do a problem, then rework it through on your own. Attend class regularly; let us build a community where you can feel good about learning, asking for help, and giving back!

"Time held me green and dying/Though I sang in my CHAINS like the sea."
— Dylan Thomas

No calculators are allowed to be used for exams.

Instructions: Books, extra papers, or calculators are not permitted. Do **all** work in the blue book(s), including scratch work; don't tear any portion of the blue book(s); keep this page. Please be neat, write the problem number, circle your answer, and explain your steps (it helps us give partial credit when you go wrong somewhere). You have 50 minutes to complete the exam. Each problem is worth 20 points.

1. An oil drill moves up and down vertically; its position as a function of time is $s(t) = \cos t - \sin t$ (where $s = 0$ is ground level).

How high above ground is the drill at $t = \dfrac{7}{4}\pi$?

What are the *speed* and the acceleration at $t = \dfrac{\pi}{2}$?

At what times in the interval $0 \le t \le 2\pi$ does the drill change direction (from up to down or vice versa)?

Requires knowledge of the applications of calculus (velocity, acceleration).

Answer:

$$s(t) = \cos t - \sin t,$$

$$s\left(\frac{7}{4}\pi\right) = \cos\left(\frac{7}{4}\pi\right) - \sin\left(\frac{7}{4}\pi\right) = \frac{\sqrt{2}}{2} - \frac{\sqrt{2}}{2} = \sqrt{2}$$

$$v(t) = s'(t) = -\sin t - \cos t,$$

$$v\left(\frac{\pi}{2}\right) = -\sin\left(\frac{\pi}{2}\right) - \cos\left(\frac{\pi}{2}\right) = -1 - 0 = -1$$

since the velocity is –1, the speed is 1.

$$a(t) = v'(t) = -\cos t + \sin t,$$

$$a\left(\frac{\pi}{2}\right) = -\cos\left(\frac{\pi}{2}\right) + \sin\left(\frac{\pi}{2}\right) = -0 + 1 = 1$$

In the interval $0 \le t \le 2\pi$, $v(t) = 0 \Rightarrow -\sin t - \cos t = 0$, and since this cannot happen when $\cos t = 0$, it happens if and only if $\dfrac{\sin t}{\cos t} = \tan t = -1$, namely when $\dfrac{3}{4}\pi$ or $\dfrac{7}{4}\pi$.

Interval of t	$\left[0, \dfrac{3}{4}\pi\right)$	$\left(\dfrac{3}{4}\pi, \dfrac{7}{4}\pi\right)$	$\left(\dfrac{7}{4}\pi, 2\pi\right)$
Sign of v	-	+	-

Thus at $\dfrac{3}{4}\pi$ or $\dfrac{7}{4}\pi$ the drill changes direction from down to up. At $t = \dfrac{7}{4}\pi$, the drill changes direction from up to down.

2. Find the derivative of the following functions (on their understood domain):

$$f(x) = \cos\left(x^{\frac{1}{3}}\right) \qquad g(x) = (\cos x)^{\frac{1}{3}}$$

$$h(x) = \frac{x + 2}{\sin x} \qquad k(x) = x^{-2} \sin\left(\frac{\pi}{2}\right)$$

This problem tests differentiation skills. It is best to apply the rules neatly: for example, chain and quotient rule, showing each step, but not quoting each theorem.

Answer:

$$f'(x) = -\sin\left(x^{1/3}\right)\left(x^{1/3}\right)' = -\frac{1}{3}\sin\left(x^{1/3}\right)x^{-2/3}$$

$$g'(x) = \frac{1}{3}(\cos x)^{-2/3}(\cos x)' = -\frac{1}{3}(\cos x)^{-2/3}\sin x.$$

$$h'(x) = \frac{(x + 2)'\sin x - (x + 2)(\sin x)'}{(\sin x)^2} = \frac{\sin x - (x + 2)(\cos x)}{(\sin x)^2}$$

Since $\sin\dfrac{\pi}{2} = 1, k(x) = x^{-2}$ and $k'(x) = -\dfrac{2}{x^3}$.

Problems 3 and 5 are word problems, where the data must be given mathematical labels and the answer has to be justified by stating and applying the appropriate theory—implicit differentiation and max/mm, respectively.

3. A truck is being dragged toward a building by a rope fastened to a pulley, which is at a height of 30 meters on the wall. The rope is being wound at a rate of 4 meters per minute. How fast is the truck approaching when it is 40 meters from the building?

rope   pulley

truck   →   bldg

Answer:

Let $x(t)$ be the horizontal distance from the truck to the building and $s(t)$ be the length of the rope at t. Then,

$$s(t)^2 = x(t)^2 + 30^2 \Rightarrow$$

$$2s(t)\frac{ds}{dt} = 2x(t)\frac{dx}{dt} \Rightarrow$$

$$\frac{dx}{dt} = \frac{s(t)}{x(t)}\frac{ds}{dt}$$

From the data, $\dfrac{ds}{dt} = -4 \text{ m/min}$, and when $x = 40$, $s = \sqrt{40^2 + 50^2} = 50 \text{ m}$.

Therefore, $\dfrac{dx}{dt} = \dfrac{50}{40} \cdot (-4) = -5$.

The truck is approaching at a rate of 5 m/min.

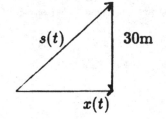

4. Find the slope of the tangent line to the curve $\sin(xy) = \dfrac{\sqrt{2}}{2}$ at the point $(1/4, \pi)$.

> This is a geometric problem. You must know how (implicit) differentiation relates to the slope of the curve.

Answer:

$$\sin(xy) = \frac{\sqrt{2}}{2} \Rightarrow \cos(xy)\left(y + x\frac{dy}{dx}\right) = 0.$$

When $x = \dfrac{1}{4}$ and $u = \pi$, $\cos\dfrac{\pi}{4} = \dfrac{\sqrt{2}}{2} \neq 0$, so $y + x\dfrac{dy}{dx} = -\dfrac{y}{x}$, as $x \neq 0$.

Thus the slope of the tangent line at $\left(\dfrac{1}{4}, \pi\right)$ is -4π.

5. A piece of sheet metal is rectangular, 5 feet wide and 8 feet long. Equal squares are to be cut from its corners and the resulting piece of metal folded and welded to form a box with an open top. What are the dimensions of the box of largest possible volume?

Answer:

If we call the dimensions of the box h, l, w for height, length, and width, respectively, we have:

$$2h + w = 5 \Rightarrow w = 5 - 2h,$$
$$2h + l = 8 \Rightarrow l = 8 - 2h$$

so the volume V equals

$$(5 - 2h)(8 - 2h) \text{ and } \frac{dV}{dh} = 12h^2 - 52h + 40 = 4(3h - 10)(h - 1)$$

The critical numbers are $h = \dfrac{10}{3}$ and 1, but the possible range of value of h is

$0 \le h \le \dfrac{5}{2}$, so only $h = 1$ falls within it. At the endpoints, $V = 0$, so the volume of

the box is a max of 18 ft^3 when $h = 1$, $w = 3$, $l = 6$.

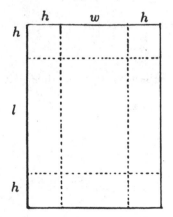

Bad is FINAL in this light.—Wallace Stevens

Instructions: Books, extra papers, or calculators are not permitted. Do **all** work in the blue book(s) including scratch work; don't tear any portion of the blue book(s); keep this page. Please be neat, explain your steps: it helps me give partial credit; the bare answer without work is not acceptable. You have 2 hours to complete your work. There are 10 problems for a total of 200 points.

1. (25 points)
   (a) Evaluate the following limits when they exist or explain why they do not exist:
   $$\lim_{x \to 0} x \sin\left(\frac{2}{x}\right)$$

   > This problem tests understanding of limits and continuity. In teaching these concepts, I stress justification by pictures.

Answer:

$$\lim_{x \to 0} x \sin\left(\frac{2}{x}\right) = 0 \text{ since } \left|\sin\left(\frac{2}{x}\right)\right| \le 1.$$

$$\lim_{x \to \frac{\pi}{2}} x \tan x \text{ does not exist since } \lim_{x \to \frac{\pi}{2}} x = \frac{\pi}{2} \text{ but } \lim_{x \to \frac{\pi}{2}^{\pm}} \tan x = \mp\infty.$$

   (b) If possible, find the number(s) c such that the following function is continuous at $x = 0$:
   $$f(x) = \begin{cases} 2x + c & x \le 0 \\ \dfrac{\sin x}{x} & 0 < x \end{cases}$$

Answer:

$$\lim_{x \to 0-} (2x + c) = \lim_{x \to 0+} \frac{\sin x}{x} \Rightarrow c = 1$$

2. (20 points)

(a) For which xs is the following function differentiable? (Explain but do not give a rigorous proof.)

$$f(x) = |x^2 - 3|$$

Compare this to problems 2 and 4 in the midterm. It tests the same skills. In addition, here the concept of differentiability is tested—a somewhat delicate property, the understanding of which I stress by picture.

Answer:

$$f(x) = |x^2 - 3| = \left|\left(x + \sqrt{3}\right)\left(x - \sqrt{3}\right)\right|$$

is differentiable everywhere except for $x = \pm\sqrt{3}$ because the graph has a cusp at these points.

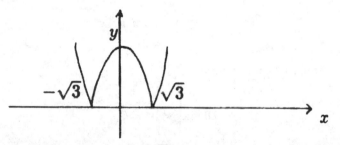

More precisely, the function coincides with $\pm(x^2 - 3)$ as $|x - \sqrt{3}| > 0$, respectively < 0, and a polynomial is everywhere differentiable. But the limits of the derivative of $|x^2 - 3|$ as $x \to \sqrt{3}^{\pm}$, as well as $x \to -\sqrt{3}^{\pm}$, do not coincide.

(b) Find an equation for the normal line to the graph of the following function when $x = 1$:

$$g(x) = \frac{1}{x + 1}$$

Answer:

$$g(x) = \frac{1}{x + 1} \Rightarrow g'(x) = \frac{1}{(x + 1)^{2}}$$

$g(1) = \ , g'(1) = -\ ,$ the slope of the normal at $x = 1$ is $m = 4$, so the normal line is given by

$$y = 4(x - 1) +$$

(c) Find *dy/dx*:

$$y = \cos(x^3 + 4 - x)$$

Answer:

$$dy/dx = -\sin(x^3 + 4 - x) \cdot (3x^2 - 1)$$

Problems 3 and 4 are similar to problems 3 and 5 in the midterm.

3. (20 points) It's a very hot day. Ice cream is leaking from the bottom of an ice cream cone at a rate of 4 cm³/min. If the wafer that contains the ice cream is a cone of radius 5 cm and height 10 cm, how fast is the level of the ice cream decreasing when the ice cream is 4 cm deep?

Answer:

$$\frac{dV}{dt} = 4 \text{ cm}^3 / \text{min}$$

Work shown:

Let h = depth, r = radius at time t.

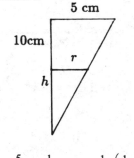

$$V = \frac{1}{3}\pi r^2 h, \quad r = \frac{5}{10}h = \frac{1}{2}h \Rightarrow V = \frac{1}{3}\pi\left(\frac{1}{2}h\right)^2 h = \frac{1}{12}\pi h^3$$

$$\frac{dV}{dt} = \frac{1}{4}\pi h^2 \frac{dh}{dt} \Rightarrow \frac{dh}{dt} = \frac{dV}{dt} \cdot 4 \cdot \frac{1}{\pi h^2} = \frac{1}{\pi} \text{ cm} / \text{min}$$

4. (20 points) A house is located 500 meters from a road, and there is a 500–m path that goes directly from the door D of the house to a point R on the road. The bus stop B is 300 m away from R down the road. If I can walk on the path at 4 m per second and on the grass that surrounds it at 1 m per second, at what point P on the path do I want to step into the grass so that the route $DP + PB$ takes me from the house to the bus in the least possible time?

Answer:

$$10\sqrt{60} \text{ m from the road}$$

Let $RP = x$.

Then $\sqrt{x^2 + 300^2} = \sqrt{x^2 + 90{,}000}$.

To minimize $T = \dfrac{500 - x}{4} + \dfrac{\sqrt{x^2 + 90{,}000}}{1}$, we set $T' = 1 \cdot + (x^2 + 90{,}000)^{-1/2} \cdot 2x = 0$.

This happens when $\dfrac{x}{\sqrt{x^2 + 90{,}000}} = \dfrac{1}{4}$, which implies

$(4x)^2 = x^2 + 90{,}000$ and $x = 10\sqrt{60}$

5. (30 points) For the following function, discuss the domain, continuity, horizontal and vertical asymptotes, extrema, whether increasing/decreasing, concave up/down, and sketch the graph:

$$f(x) = \frac{1}{x^2 - 1}$$

This problem covers the very important topic of graph sketching. To succeed here, the student ought to follow the strategy outlined in our textbook: analyze behavior at infinity, max/mm, concavity/convexity.

Answer:

$$f(x) = \frac{1}{x^2 - 1} = \frac{1}{(x + 1)(x - 1)}$$

Domain: $(-\infty, -1) \cup (-1, 1) \cup (1, \infty)$.

Vertical asymptotes: $x = \pm 1$.

Continuous except at $x = \pm 1$.

$f'(x) = \dfrac{2x}{(x^2 - 1)^2} = 0$ when $x = 0$, $f(x)$ increasing for $x < 0$, decreasing for $x > 0$;

thus the only local max is $f(0) = -1$.

243

$$f''(x) = \frac{-2(x^2-1)^2 + 2x \cdot 2(x^2-1) \cdot 2x}{(x^2-1)^4} = \frac{2(3x^2+1)}{(x^2-1)^3}$$

From the sign $f''(x)$, $f(x)$ is concave up on $(-\infty,-1)$ and $(1,\infty)$, and concave down on $(-1,1)$.

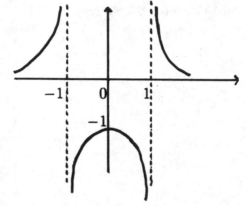

6. (10 points) Use the Fundamental Theorem of Calculus to find

$$\frac{d}{dx}f(x) \quad \text{where} \quad f(x) = \int_{x3}^{7x} \sqrt{1+t+t^2}\,dt$$

> This problem tests the very deep Fundamental Theorem of Calculus. However, most students find this so difficult to grasp that I try to train them to do this problem mechanically.

Answer:

$$\frac{d}{dx}\int_{x^3}^{7x} \sqrt{1+t+t^2}\,dt = \frac{d}{dx}\int_{x^3}^{a} \sqrt{1+t+t^2}\,dt + \frac{d}{dx}\int_{a}^{7x} \sqrt{1+t+t^2}\,dt =$$

$$-\sqrt{1+x^3+x^6}\,(3x^2) + \sqrt{1+7x+49x^2}\cdot 7.$$

> Problems 7, 8, and 9 test integration skills, but, more importantly, also test geometric understanding, so that the integrals can be set up correctly. I strongly emphasize the use of pictures.

7. (15 points) Find the area of the region bounded by the following graphs:
$$y = x^2 + x, \ y = x + 1.$$

Answer:

The given graphs meet when $x^2 + x = x + 1$, namely $x = \pm 1$.

$$A = \int_{-1}^{1} (x + 1 - x^2 - x)\,dx = \left[x - \frac{1}{3}x^3 \right]_{-1}^{1} = \frac{4}{3}$$

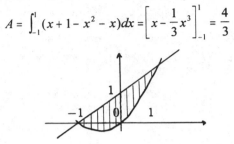

8. (15 points) Use the method of cylindrical shells to calculate the volume of the solid obtained by revolving about the y-axis the region of the first quadrant (that is, those points (x,y) with both x and y nonnegative) which is between the graph $y = \sqrt{4 - x^2}$ and the x-axis.

Answer:

To use the method of shells (around the y-axis), we need to bound the region by the graph of a function of x:

$$y = \sqrt{4 - x^2}, \quad V = \int_{0}^{2} 2\pi x \sqrt{4 - x^2}\,dx = -\pi \int_{4}^{0} u^{1/2}\,du \ \text{(substitution: } u = 4 - x^2 \text{)}$$

$$= \pi \left[\frac{2}{3} u^{3/2} \right]_{0}^{4} = \frac{16}{3}\pi$$

9. (30 points)

(a) Find the length of the arc determined by the equation: $9y^2 = 4(x + 1)^3$ for $0 \le x \le 1$.

(b) Find the surface area of the solid generated by revolving about the x-axis the region bounded by the graph $2\sqrt{x}$ and the x-axis for $1 \le x \le 2$.

Answer:

(a) $$y = \frac{2}{3}(x + 1)^{3/2}, \ y' = (x + 1)^{1/2}, \ (y')^2 = x + 1.$$

$$L = \int_0^1 \sqrt{1 + x + 1}\,dx = \int_2^3 u^{1/2}du \ \text{(substitution: } u = 2 + x) = \left[\frac{2}{3}u^{3/2}\right]_2^3 = \frac{2}{3}(3^{3/2} - 2^{3/2})$$

(b) $$y = 2\sqrt{x} \quad y' = x^{-1/2}, \quad (y')^2 = \frac{1}{x}.$$

$$S = \int_1^2 2\pi(2\sqrt{x})\sqrt{1 + \frac{1}{x}}\,dx = 4\pi\int_1^2 \sqrt{x + 1}\,dx = 4\pi\int_2^3 u^{1/2}du \ \text{(substitution: } u = x + 1)$$

$$= 4\pi\left[\frac{2}{3}u^{3/2}\right]_2^3 = \frac{8}{3}\pi(3^{3/2} - 2^{3/2}).$$

10. (15 points) If a particle is moving on a straight line with velocity $v(t) - 3t + 2$, find for which times the particle is at rest, the distance traveled from time $t = 0$ to $t = 2$, and also the displacement for the same time interval.

Compare this problem to problem 1 in the midterm.

Answer:

$$v(t) = t^2 - 3t + 2 = (t - 2)(t - 1) = 0 \text{ at } t = 1,2.$$

Distance:

$$\int_0^1 (t^2 - 3t + 2)dt - \int_1^2 (t^2 - 3t + 2)dt = \left[\frac{t^2}{3} - \frac{3}{2}t^2 + 2t\right]_0^1 - \left[\frac{t^2}{3} - \frac{3}{2}t^2 + 2t\right]_1^2 = 1$$

Displacement:

$$\int_0^2 (t^2 - 3t + 2)dt = \left[\frac{t^2}{3} - \frac{3}{2}t^2 + 2t\right]_0^2 = \frac{2}{3}$$

CHAPTER 21

UNIVERSITY OF CHICAGO

MATHEMATICS 152: CALCULUS

Fedor A. Andrianov, Lecturer

THIS COURSE IS PART OF A THREE-SEMESTER SEQUENCE. THREE EXAMINATIONS PER quarter are given in this course sequence (a total of nine tests per year). My primary objectives in teaching this course are:

1. To show students the beauty, the grandeur, and the power of mathematics, the Queen of Sciences.

2. To explain the genesis of mathematical abstraction, the motivation behind its principle notions, and the driving forces of its development.

3. To teach students the art of integrating precise thinking (analysis) with creative dreaming (synthesis), that is, to develop all aspects of their abstract reasoning. One of the most essential points here is to understand that a fixed amount of knowledge is never enough and that the ability to make an educated guess is equally important for problem solving.

4. To train students to use basic mathematical notions, ideas, and techniques efficiently in solving real-world problems. This includes an ability to mirror a real-life problem in a mathematical model, effective theoretical study of the particular model, and understanding of its relation to a more general calculus framework.

5. Application of derived abstract results to original problems.

As a result of taking this course students should:

1. Develop a good, solid *understanding* of basic concepts, results, and techniques

of modern calculus. These include the notion of limit of a function or a sequence, continuity, and differentiability, the Mean-Value Theorem, maxima and minima, integration and the Newton-Leibniz formula, solution of differential equations, and approximation techniques. (The key word here is *UNDERSTANDING,* as opposed to formal memorizing. In particular, students should not only learn the basic notions, but also understand their interconnections and real-world motivations.)

2. Acquire confidence in standard computational and applied exercises (similar to those presented in the exams that follow).

3. Be able to do such theoretical reasoning as epsilon-delta proofs of continuity of elementary functions or derivation of particular corollaries from the main theorems.

The course examinations test both key skills that students are supposed to acquire: understanding of main concepts (theory) and proficiency in computational exercises (practice). This is clearly reflected in the choice of test problems: some of them are purely computational; some require knowledge of main theoretical results and an ability to apply them to a particular situation; finally, some problems just test understanding of basic notions.

A common mistake students make in a calculus course is putting stress on the formal memorization of isolated words (not ideas) and isolated techniques. Understanding of the general picture is often lacking, which results in inability to solve problems that are even slightly different from those detailed in class. My advice is to put the stress on *understanding the material.* Calculus is not poetry!

MIDTERM EXAM

It is relatively easy to outline what makes an answer a good one. It should be rather self-evident that a successful response MUST BE complete (i.e., explicitly include all key steps of solution) and well organized (i.e., follow the natural flow of ideas leading to the final result). Since the main point of studying a subject is still to learn something about it (and not just to get a grade), it should be clear that mastery of course factual material, understanding of its major points, and good problem-solving skills are absolute MUSTS, at least on an average level. These are prerequisites to any successful response on any test and are presupposed automatically (at least by a teacher).

Advice applicable to any mathematical problem: "PLEASE, JUST THINK ABOUT IT!"

1. (10 points): Compare the following expressions:

$$\frac{d}{dx}\left(\int_a^x f(t)\,dt\right), \quad \int_a^x \frac{d}{dt}\big(f(t)\,dt\big)$$

Answer:

Set

$$F(x) = \int_a^x f(t)dt,$$

then $F(x)$ is an antiderivative for $f(x)$ and, using Newton-Leibniz formula, we see that

$$\frac{d}{dx}\left(\int_a^x f(t)dt\right) = \frac{d}{dx}\big(F(x) - F(a)\big) = f(x).$$

On the other hand,

$$\int_a^x \frac{d}{dt}\big(f(t)\big)dt = \int_a^x d\big(f(t)\big) = f(t)\Big|_a^x = f(x) - f(a)$$

Thus, the expressions differ by a constant $f(a)$.

> **Successful response requires:**
> 1. Understanding that integration and differentiation are operations inverse to each other
> 2. Knowing that $f(x)$ as just defined (by an integral) gives an example of an antiderivative for $f(x)$
> 3. Familiarity with the Fundamental Theorem of Integral Calculus (i.e., with the Newton-Leibniz formula)

2. (15 points): Find the critical points, classify the extreme values, and sketch the graph of the function

$$F(x) = \int_0^x t(t-3)^2 \, dt$$

Answer:

From the definition of $f(x)$ we see that $f'(x) = x(x - 3)^2$ and then $f''(x) = 3(x-1)(x-3)$. Both $f'(x)$ and $f''(x)$ always exist; therefore, the only critical points of $f(x)$ are points where $f'(x)$ is zero, that is, $x = 0$ or $x = 3$, and the only candidates for inflection points are points where $f''(x)$ is zero, that is, $x = 1$ or $x = 3$.

Examining the sign of $f'(x)$ and $f''(x)$ on both sides of these points, we arrive at the following table:

x	$(-\infty,0)$	0	$(0,1)$	1	$(1,3)$	3	$(3,+\infty)$
F'	$-$	0	$+$	$+$	$+$	0	$+$
F''	$+$	$+$	$+$	0	$-$	0	$+$
F	$\downarrow$	0, Min.	$\uparrow$	Infl. p.	$\uparrow$	Infl. p.	$\uparrow$

which classifies critical points, points of inflection, and extrema of f.

Finally, the preceding considerations allow us to sketch the graph of $f(x)$:

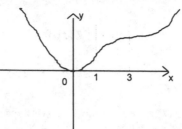

Successful response requires:
1. Recognizing that $f(x)$ is an antiderivative for $x(x-3)^2$
2. Knowing the definitions of critical points, points of inflection, and extrema, and being able to visualize images of these notions
3. Ability to differentiate simple functions
4. Ability to investigate sign of a simple function on an interval
5. Ability to sum up investigations and to visualize them in a sketch

3. (30 points): Evaluate the following integrals:

$$\int_{-\pi/2}^{\pi} |\cos x| dx, \quad \int x^{n-1} \sqrt{a + bx^n} dx, \quad \int_0^3 \frac{r}{\sqrt{r^2+16}} dr$$

Answer:

$$\int_{-\pi/2}^{\pi/2} |\cos x| dx = \int_{-\pi/2}^{\pi/2} (\cos x) dx + \int_{\pi/2}^{\pi} (-\cos x) dx = \sin x \Big|_{-\pi/2}^{\pi/2} - \sin x \Big|_{\pi/2}^{\pi} = 3.$$

Next, let $u = a + bx^n$, then $du = bnx^{n-1}dx$ and

$$\int x^{n-1} \sqrt{a + bx^n} dx = \int \frac{\sqrt{u}}{bn} du = \frac{2 \cdot u^{3/2}}{3 \cdot bn} + C = \frac{2(a + bx^n)^{3/2}}{3bn} + C.$$

Next, let $u = r^2 + 16$, then $du = 2rdr$ and

$$\int_0^3 \frac{r}{\sqrt{r^2 + 16}} dr = \int_{16}^{25} \frac{du}{2\sqrt{u}} \Big|_{16}^{25} = 5 - 4 = 1$$

Successful answer requires:
1. Knowing the definition of the Absolute Value function and understanding that integration (or differentiation) of an absolute value requires splitting the domain of integration (or differentiation) into two parts, according to the sign of expression inside the absolute value. (In the preceding example, cos x is positive on [−p/2,p/2] but negative on [p/2,p], so we accordingly split the original integral into two simpler parts not involving absolute values.)
2. Proficiency in the technique of change of variables in integrals ("u-substitution"). This includes the ability to recognize which particular u-substitution will reduce the integral in question to its simplest form and the skill to make the substitution and evaluation without errors. Here students should not forget to adjust limits of integration according to a chosen u-substitution when dealing with indefinite integrals.
3. Knowledge of the rules of integration of elementary functions.

4. (25 points): Find the average distance of the parabolic arc $y = x^2, \quad x \in \left[0, \sqrt{3}\right]$ from the origin.

Answer:

The distance of a point (x_0, y_0) of the parabolic arc $y = x^2$ from the origin $(0,0)$ is given by $D(x_0) = \sqrt{x_0^2 + y_0^2} = \sqrt{x_0^2 + x_0^4}$.

Therefore, the average distance on the interval is given by

$$D_{\text{avg}} = \frac{1}{\sqrt{3} - 0} \int_0^{\sqrt{3}} \sqrt{x^2 + x^4}\, dx = \frac{1}{\sqrt{3}} \int_0^{\sqrt{3}} \sqrt{1 + x^2}\, dx = \frac{1}{2\sqrt{3}} \int_0^{\sqrt{3}} \sqrt{1 + x^2}\, d\left(1 + x^2\right) =$$

$$= \frac{1}{2\sqrt{3}} \int_1^4 \sqrt{u}\, du = \frac{2 \cdot u^{3/2}}{3 \cdot 2\sqrt{3}}\bigg|_1^4 = \frac{4^{3/2} - 1}{3\sqrt{3}} = \frac{7}{3\sqrt{3}}.$$

Successful answer requires:
1. Knowing the definition of the average of a function on an interval.
2. Proficiency in the technique of change of variables in integrals ("u-substitution"). This includes the ability to recognize which particular u-substitution will reduce the integral in question to its simplest form and the skill to make the substitution and evaluation without errors. Here students should not forget to adjust limits of integration according to a chosen u-substitution when dealing with indefinite integrals.
3. Knowledge of the rules of integration of elementary functions.

5. (20 points): Sketch the region Ω bounded by the curves

$$F(x) = \int_0^x t(t - 3)^2\, dt$$

and find the volume of the solid generated by revolving this region about the x-axis.

Answer:

The sketch of the region is:

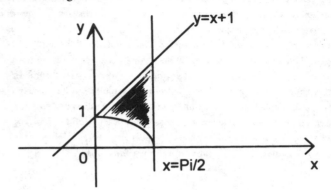

Then the volume of the solid generated by revolving the region around x-axis is given by

$$V = \int_0^{\pi/2} \pi\left((x+1)^2 - \cos^2 x\right)dx = \pi \int_0^{\pi/2} x^2 + 2x + 1 - 0.5 - 0.5 \cdot \cos(2x)dx =$$

$$= \pi\left(\frac{x^3}{3} + x^2 + \frac{x}{2} - \frac{\sin(2x)}{4}\right)\Bigg|_0^{\pi/2} = \frac{\pi^4}{24} + \frac{\pi^3}{4} + \frac{\pi^2}{4}$$

Successful answer requires:
  1. Knowing the formulas expressing a volume of revolution as an integral
  2. Ability to evaluate a simple integral
  The only trouble point in this example is the necessity of using the trigonometric identity $2\cos^2 x = 1 + \cos(2x)$ in the process of evaluation of the integral.
  The identity is rather special (isolated), and a teacher should not require students to memorize it (never!). If necessary, this identity can be given to the class as a hint for the problem.

1. (5 points): Give the definition of an antiderivative for a continuous function $f(x)$ on the interval $[a,b]$.

Answer:

A continuous function $f(x)$ on $[a,b]$ is called an antiderivative for $f(x)$ on $[a,b]$ if and only if $f'(x) = f(x)$ for all $x \in (a,b)$.

Successful answer requires knowing the characteristic properties of an antiderivative. The essential property of an antiderivative that is often overlooked by students is its continuity.

2. (5 points): State the Fundamental Theorem of Integral Calculus.

Answer:

If $f(x)$ is continuous on $[a,b]$ and $f(x)$ is any antiderivative of $f(x)$ on $[a,b]$, then

$$\int_a^b f(x)dx = F(b) - F(a).$$

A successful answer requires close attention to all details of the statement of the theorem. It is essential to mention that $f(x)$ should be continuous and that the Newton-Leibniz formula holds for an *arbitrary* antiderivative of $f(x)$.

3. (10 points): Show that, if $f(x)$ is continuous on $[a,b]$, then there exists at least one number $c \in [a,b]$ for which

$$\int_a^b f(x)dx = f(c) \cdot (b-a)$$

Answer:

This result is a corollary of the Mean Value Theorem. In fact, let $f(x)$ be an antiderivative of $f(x)$ on $[a,b]$. In particular $f(x)$ is continuous on $[a,b]$ and differentiable on (a,b). Then, according to the Mean Value Theorem, there exists $c \in [a,b]$ such that

$$F(b) - F(a) = F'(c) \cdot (b-a),$$

which is equivalent to

$$\int_a^b f(x)dx = f(c) \cdot (b-a),$$

since $f'(c) = f(c)$. End of proof.

Successful answer requires thoughtful combination of the Mean Value Theorem and the Newton-Leibniz formula. A heuristic argument can be as follows: we want to connect value of the integral with value of the function $f(x)$. The integral is essentially an antiderivative (via the Newton-Leibniz formula), so we want to connect values of an antiderivative with the value of the function. Switching our attention to the antiderivative (as another function), we can say that we want to connect values of a differentiable function with the value of its derivative. An example of such a connection is given by the Mean Value Theorem. Formalization of this argument leads to the preceding solution. Regarding a rigorous solution itself, it is essential to mention such properties of an antiderivative as continuity and differentiability on appropriate intervals, since, without these properties, the Mean Value Theorem would not be applicable.

In some situations a teacher may want to simplify the task by explicitly suggesting use of the Mean Value Theorem from the start (as a hint for the problem).

4. (10 points): Calculate the derivative

$$\frac{d}{dx} \int_{3x}^{1/x} (\cos 2t)\, dt$$

Answer:

Consider the following:

$$\frac{d}{dx}\left(\int_{3x}^{1/x} \cos(2t)\, dt \right) = \frac{d}{dx}\left(\int_{c}^{1/x} \cos(2t)\, dt - \int_{c}^{3x} \cos(2t)\, dt \right) =$$

$$= -\cos\frac{2}{x} \cdot x^{-2} - \cos(6x) \cdot 3 = -\left(x^{-2} \cdot \cos\frac{2}{x} + 3 \cdot \cos(6x) \right)$$

Successful answer requires:
1. Understanding that integration and differentiation are operations inverse to each other.
2. Knowing that an integral with a variable upper limit gives an example of an antiderivative.
3. Recalling additivity of integrals. This property allows reduction of the original integral to a difference of two integrals with variable upper and constant lower limits that can be easily differentiated.
4. Using the chain rule for differentiation of composite functions.

5. (15 points): Let $f(x) = \int_1^{2x} \sqrt{16 + t^4}\, dt$

Prove that $f(x)$ has an inverse and find $(f^{-1})'(0)$.

Answer:

Since

$$f'(x) = \frac{d}{dx} \int_1^{2x} \sqrt{16 + t^4}\, dt = \sqrt{16 + (2x)^4} \cdot 2 > 0, \forall x,$$

then $f(x)$ is a 1-to-1 (i.e., injective) function and therefore it has an inverse. Next, if $f(x_0) = 0$ then

$$\left(f^{-1}\right)'(0) = \frac{1}{f'(x_0)} = \frac{1}{2\sqrt{16 + (2x_0)^4}}$$

and it remains to find x_0. We have

$$f(x_0) = 0 \Leftrightarrow \int_1^{2x_0} \sqrt{16 + t^4}\, dt = 0 \Leftrightarrow 1 = 2x_0 \Leftrightarrow x_0 = \frac{1}{2},$$

and so

$$\left(f^{-1}\right)'(0) = \frac{1}{2\sqrt{16 + (2 \cdot 1/2)^4}} = \frac{1}{2\sqrt{17}}.$$

Successful answer requires:
1. Understanding that integration and differentiation are operations inverse to each other
2. Knowing that an integral with a variable upper limit gives an example of an antiderivative
3. Using the chain rule for differentiation of composite functions
4. Understanding that, in order to have an inverse, a function must be injective
5. Knowing basic tests for injectivity of a continuous function (such as a constant sign of its derivative)
6. Knowing how to express derivative of an inverse function via derivative of original function
7. Understanding that a definite integral of a positive function can be zero only if the interval of integration is zero

6. (15 points): Calculate the following integrals:

$$\int \frac{\sin x}{2 + \cos x} dx, \quad \int \frac{\sin(e^{-2x})}{e^{2x}} dx, \quad \int \frac{\log_2 x^3}{x} dx$$

Answer:

Let $u = 2 + \cos x$, then $du = -\sin x \, dx$ and

$$\int \frac{\sin x}{2 + \cos x} dx = -\int \frac{du}{u} = \ln|u| + C = -\ln|2 + \cos x| + C$$

Next, let $u = e^{-2x}$, then $du = e^{-2x}(-2)dx$ and

$$\int \frac{\sin(e^{-2x})}{e^{2x}} dx = -\frac{1}{2} \int (\sin u) du = \frac{-\cos u}{-2} + C = \frac{\cos(e^{-2x})}{2} \cdot C$$

Next, let $u = \ln x$, then $du = dx / x$ and

$$\int \frac{\log_2 x^3}{x} dx = \frac{3}{\ln 2} \int \frac{\ln x}{x} dx = \frac{3}{\ln 2} \int u \, du = \frac{3(\ln x)^2}{2 \ln 2} + C$$

> Successful answer requires proficiency in the technique of change of variables in integrals ("u-substitution"). This includes the ability to recognize which particular u-substitution will reduce the integral in question to its simplest form and the skill to make the substitution and evaluation without errors. Here the student should not forget to adjust the limits of integration according to a chosen u-substitution when dealing with indefinite integrals. Also required is knowledge of the rules of integration of elementary functions.

7. (15 points): An object starts from rest at the point x_0 and moves along the x-axis with constant acceleration a. Derive the formula for the velocity and position of the object at any time $t \geq 0$. Show that the average velocity over any time interval $[t_1, t_2]$ is the average of the initial and final velocities on that interval.

Answer:

Since $a = v'(t)$, then

$$v(t) = \int a \, dt = at + v(0),$$

but since the object starts from rest, that is, $v(0) = 0$, then
$$v(t) = at.$$

Next, since $v(t) = x'(t)$, then

$$x(t) = \int v(t) dt = \int (at) dt = \frac{at^2}{2} + x(0),$$

but $x(0) = x_0$, so $x(t) = (at^2)/2 + x_0$.

Finally, if $[t_1, t_2]$ is a time interval, then

$$v_{avg} = \frac{1}{t_2 - t_1} \int_{t_1}^{t_2} v(t)dt = \frac{at_2^2 - at_1^2}{2(t_2 - t_1)} = \frac{at_2 + at_1}{2} = \frac{v(t_1) + v(t_2)}{2},$$

which proves that the average velocity over any time interval is the average of the initial and final velocities on that interval.

Successful answer requires:
1. Knowing that velocity (acceleration), being the rate of change of position (of velocity), can be expressed as derivative of position (of velocity) with respect to time
2. Knowing that, to reconstruct a function from its derivative, indefinite integrals are used
3. Being aware that an indefinite integral is in fact a *family* of functions and that a particular choice is made according to initial conditions
4. Knowing the definition of the average of a function on an intervaL.

8. (15 points): In a bacteria-growing experiment, a biologist observes that the number of bacteria in a certain culture triples every 4 hours. After 12 hours, it is estimated that there are 1 million bacteria in the culture. How many bacteria were present initially and what is the doubling time for the bacteria population?

Answer:

Since bacteria population $P(t)$ increases by the same factor during periods of the same duration, it satisfies the differential equation $P'(t) = kP(t)$, which implies that

$$\frac{P'(t)}{P(t)} = k \Rightarrow \int \frac{P'(t)}{P(t)} dt = kt + C \Rightarrow \ln|P(t)| = kt + C.$$

Thus, the bacteria population changes according to the following law:

$$P(t) = P(0) \cdot e^{k \cdot t}.$$

Next,

$$3 = \frac{P(t+4)}{P(t)} = e^{4 \cdot k} \Rightarrow k = \frac{\ln 3}{4}$$

and

$$10^6 = P(12) = P(0) \cdot e^{12 \cdot k} \Rightarrow P(0) = \frac{10^6}{e^{3 \cdot \ln 3}} = \frac{10^6}{27}.$$

Finally, let us find the doubling time Δt:

$$2 = \frac{P(t + \Delta t)}{P(t)} = e^{k \cdot \Delta t} \Rightarrow \Delta t = \frac{4 \cdot \ln 2}{\ln 3}.$$

Successful answer requires:
1. Basic knowledge about exponential growth and decay including what type of experimental data suggests exponential law and what differential equation reflects this law
2. Ability to solve a simple differential equation via the separation of variables method
3. Awareness that an indefinite integral is in fact a *family* of functions and that a particular choice is made according to initial conditions
4. Knowing the definition of *doubling time*

9. (5 points): Give the definition of the Euler number *e*.

Answer:

The Euler number *e* is uniquely defined by the following equality:

$$\ln e = \int_1^e \frac{dt}{t} = 1$$

Here the student just needs to know the definition—nothing else.

10. (5 points): Give an estimate from above and an estimate from below for the Euler number *e*. Please, JUSTIFY your answer!

This problem has infinitely many right answers and is designed to give students an opportunity to demonstrate their ingenuity as well as their knowledge. A possible answer follows.

Answer:

Note that

$$\ln x = \int_1^x \frac{dt}{t}, \forall x \in (0, +\infty) \Rightarrow \ln' x = \frac{1}{x} > 0, \forall x \in (0, +\infty).$$

In particular $\ln x$ is an increasing function. Therefore

$$\ln 1 = 0 < 1 = \ln e \Rightarrow 1 < e.$$

On the other hand,

$$\ln\left(1+\frac{1}{n}\right) = \int_{1}^{1+1/n} \frac{dt}{t} > \int_{1}^{1+1/n} \frac{dt}{1+1/n} = \frac{1}{n}\cdot\frac{n}{n+1} = \frac{1}{n+1} \Rightarrow (n+1)\cdot\ln\left(1+\frac{1}{n}\right) > 1 \Rightarrow$$

$$e < \left(1+\frac{1}{n}\right)^{n+1}, \forall n.$$

which gives us estimates for *e* from above and from below.

A successful answer requires good understanding of the definition of the Euler number via the integral and ability to estimate integrals from above or from below using Riemann sums. This leads to many different estimates of *e*, depending on the chosen method. Of course, a calculator has nothing to do with this problem!

CHAPTER 22

UNIVERSITY OF TENNESSEE

MATH 125: BASIC CALCULUS

Jennifer C. Stevens, Instructor

FOUR EXAMS—THREE MIDTERMS AND A FINAL—ARE GIVEN IN THIS COURSE.
Math 125 is designed to introduce and explore the calculus of algebraic, exponential, and logarithmic functions. The objective of the course is to familiarize the student with the basic concepts and techniques of differential and integral calculus and their applications in problem solving.

An intuitive understanding of concepts is stressed over theory and rigorous proofs. The graphing calculator is used to help the students think about the geometric and numerical meaning of calculus and to approximate numerical solutions to realistic application problems.

Topics include:

◆ Limits

◆ Continuity

◆ Derivatives

◆ Techniques of differentiation

◆ Marginal analysis

◆ Curve sketching and optimization

◆ Definite and indefinite integration

◆ Integration by substitution

◆ The Fundamental Theorem of Calculus

◆ Area between curves

◆ Applications of the integral

◆ Numerical integration

All topics in this list and the objectives mentioned are tested in the final exam, with the possible exception of continuity. The second midterm exam (included here) tested students on techniques of differentiation, marginal analysis, and curve sketching.

My number one suggestion for success in this course is to attend class regularly. Try not to miss a single class. Do the assigned homework nightly. Ask questions in class. Also ask your teacher for extra help outside of class, if necessary. Find a group of students to study with. Use the chapter review problems as practice exam problems. Go back and review homework problems and quizzes.

MIDTERM EXAM

Show all work to receive full credit.

1. Find the derivative.

$$f(x) = 5x^6 + \frac{3}{x^2} - 2e^x$$

Answer:

Rewrite in power form.

$$f(x) = 5x^6 + 3x^{-2} + 2e^x$$

Differentiate, using the power rule.

$$f'(x) = 5(6x^5) + 3(-2x^{-3}) - 2e^x$$

Simplify:

$$f'(x) = 30x^5 - \frac{6}{x^3} - 2e^x$$

2. Find $g'(x)$

$$g(x) = \frac{1}{3}\ln x - \sqrt{x} + 8$$

Answer:

Rewrite in power form, then differentiate and simplify.

$$g(x) = \frac{1}{3}\ln x - x^{\frac{1}{2}} + 8$$

$$g'(x) = \frac{1}{3}\left(\frac{1}{x}\right) - \frac{1}{2}x^{-\frac{1}{2}} + 0$$

$$g'(x) = \frac{1}{3x} - \frac{1}{2\sqrt{x}}$$

3. Find $\frac{dy}{dx}$. $y = x^2\ln x$

Answer:

Differentiate, using the product rule.

$$\frac{dy}{dx} = x^2\left(\frac{1}{x}\right) + 2x\ln x$$

Simplify:

$$\frac{dy}{dx} = x + 2x\ln x$$

4. Find $f'(t)$

$$f(t) = \frac{t^2 + 4}{3t^2 + 1}$$

Answer:

Differentiate, using the quotient rule.

$$f'(t) = \frac{(3t^2 + 1)(2t) - (t^2 + 4)(6t)}{(3t^2 + 1)^2}$$

Simplify:

$$f'(t) = \frac{-22t}{(3t^2 + 1)^2}$$

5. Find the derivative of $y = (3x^5 + 2)^{10}$

Answer:

Differentiate, using the chain rule.

$$y' = 10(3x^5 + 2)^9(15x^4)$$

Simplify:

$$y' = 150x^4(3x^5 + 2)^9$$

6. Find $\dfrac{d}{dx}\left(e^{7x^2+3}\right)$

Answer:

Differentiate, using the chain rule.

$$e^{7x^2+3}(14x)$$

Simplify:

$$14xe^{7x^2+3}$$

7. Find the derivative of $y = \ln(5x^3 + 7x - 11)$

Answer:

Differentiate, using the chain rule.

$$y' = \frac{1}{5x^3 + 7x - 11}(15x^2 + 7)$$

Simplify:

$$y' = \frac{15x^2 + 7}{5x^3 + 7x - 11}$$

8. Find $R'(10)$ if $R(x) = x^3 e^{0.3x}$

Answer:

Differentiate, using the product and chain rules.

$$R'(x) = x^3 e^{0.3x}(0.3) + e^{0.3x}(3x^2)$$

Now simplify and then evaluate.

$$R'(x) = x^2 e^{0.3x}(0.3x + 3)$$
$$R'(10) = 10^2 e^{0.3(10)}(0.3(10) + 3) = 12051$$

9. If the demand equation is given by $p = -0.4x + 30$, find the marginal revenue equation.

Answer:

First find the revenue equation.

$$R(x) = xp = x(-0.4x + 30)$$

Simplify:

$$R(x) = -0.4x^2 + 30x$$

Marginal revenue is the derivative.

$$R'(x) = -0.8x + 30$$

10. If $C(x) = 5 + 2x + 0.03x^2$ is a cost equation, where $C(x)$ is the cost in dollars of x items, use calculus to estimate the cost of the 51st item.

Answer:

Find the marginal cost equation:

$$C'(x) = 2 + 0.06x$$

The approximate cost of the 51st item:

$$C'(50) = 2 + 0.06(50) = 5$$

The 51st item costs approximately $5.

11. Let $f(x) = 3x^5 - 20x^3 + 5$

a) Find all critical values.

Answer:

Differentiate:

$$f'(x) = 15x^4 - 60x^2$$

Factor and solve $f'(x) = 0$ to find critical values:

$$0 = 15x^2(x-2)(x+2)$$

Critical values: $x = 0, 2, -2$.

b) Find the largest open interval(s) where $f(x)$ is increasing.

The graph of $f(x)$ is increasing on open intervals where $f'(x) > 0$. Use a sign chart to show that $f'(x) > 0$ on $(-\infty, -2)$, $(2, \infty)$. Therefore, $f(x)$ is increasing on $(-\infty, -2)$, $(2, \infty)$.

Answer: $f(x)$ is increasing on $(-\infty, -2)$, $(2, \infty)$.

c) Find all relative maxima.

Answer:

Since -2 is a critical value and $f(x)$ is increasing on $(-\infty, -2)$ and decreasing on $(-2, 0)$, $f(-2) = 69$ is a relative maximum.

d) Find all relative minima.

Answer:

Since 2 is a critical value and $f(x)$ is decreasing on $(0, 2)$ and increasing on $(2, \infty)$, $f(2) = -59$ is a relative minimum.

12. Let $f(x) = x^5 - 5x^4 - 17$

a) Find the largest open interval(s) where $f(x)$ is concave up.

Answer:

Find the second derivative.

$$f'(x) = 5x^4 - 20x^3, \quad f''(x) = 20x^3 - 60x^2$$

Factor and solve $f''(x) = 0$ to find possible x-values where concavity changes. $0 = 20x^2(x - 3)$.

Solutions: $x = 0, 3$.

The graph of $f(x)$ is concave up on open intervals where $f''(x) > 0$.

Use a sign chart to show that $f''(x) > 0$ on $(3, \infty)$.

Therefore, $f(x)$ is concave up on $(3, \infty)$.

b) Find all inflection points.

Answer:

The function $f(x)$ changes concavity at $x = 3$; therefore, $(3, -179)$ is an inflection point.

13–16: Match the graph of each function, *f*, *g*, *h*, and *k*, to the graph of its derivative, A, B, C, D. On each of the eight graphs, the scale = 1 on both the *x*-axis and the *y*-axis.

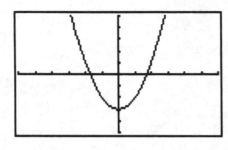

$y = f(x)$

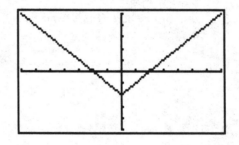

$y = g(x)$

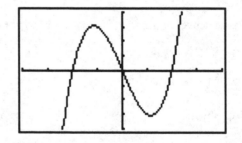

$y = h(x)$

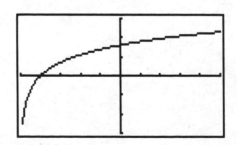

$y = k(x)$

13. The graph of $y = f'(x)$ is _____
Answer: b

14. The graph of $y = g'(x)$ is _____
Answer: d

15. The graph of $y = h'(x)$ is _____
Answer: a

16. The graph of $y = k'(x)$ is _____
Answer: c

a.

b.

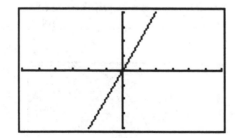

c.

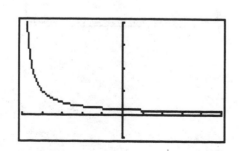

d.

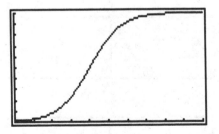

For questions 13–16: At several points on each graph, draw in a tangent line. Estimate the value and determine the sign of the slope. These slope values are the values of the derivative function. For example, in question 13, the slopes of all tangent lines of $y = f(x)$ are negative to the left of $x = 0$, and positive to the right of $x = 0$, and the slope of the tangent line at $x = 0$ is 0. The graph in B matches this description; therefore it must be the graph of $y = f'(x)$. In question 14, the slope of $y = g(x)$ is always −1 to the left of $x = 0$, and is 1 to the right of $x = 0$. Therefore, the graph of $y = g'(x)$ must be the constant function −1 to the left of $x = 0$ and the constant function 1 to the right of $x = 0$. This is the graph in D.

17. A logistic model of the U.S. population is given by

$$p(t) = \frac{500}{1 + 124e^{-0.024t}}$$

where $p(t)$ is the population of the U.S. in millions and t is the time in years with $t = 0$ corresponding to 1790. Use the graph below to estimate the year in which the inflection point occurs. What is the significance of this point in terms of the rate of growth of population? The viewing screen below is [0, 500] by [0, 500], using a scale = 50 on both axes.

Answer:

The inflection point occurs where the concavity of the graph changes. At about the fourth notch on the horizontal axis (t), the concavity of the graph changes from concave up to concave down. Since the scale = 50 and 4×50 = 200, the inflection point occurs about 200 years after the beginning of the graph. Since we are told that $t = 0$ corresponds to 1790, then the inflection point occurs in 1990. The significance of this point is that the rate of growth of population begins to decrease in 1990; population is still increasing, but by smaller and smaller amounts, beginning in 1990.

FINAL EXAM

Show all work and graphs to receive full credit.

1. Use the following graph to answer the questions. The scale on both the x- and y-axes is 1.

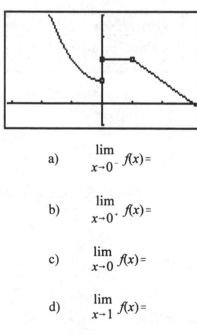

a) $\displaystyle\lim_{x\to 0^-} f(x) =$

b) $\displaystyle\lim_{x\to 0^+} f(x) =$

c) $\displaystyle\lim_{x\to 0} f(x) =$

d) $\displaystyle\lim_{x\to 1} f(x) =$

Answers:

a) 1. Follow the graph as the x-values approach 0 from the left side. The y-values approach 1.

b) 2. Follow the graph as the x-values approach 0 from the right side. The y-values approach 2.

c) Does not exist. Since the limit on one side is 1 and the limit on the other side is 2 (they are not the same number), then the limit does not exist.

d) 2. As the *x*-values approach 1 from either side, the *y*-values approach 2.

2. Fill in the table and then estimate the limit, if it exists.

$$\lim_{x \to 2} \frac{x-2}{x^2 - 5x + 6} =$$

x	1.9	1.99	1.999	2	2.001	2.01	2.1
$\dfrac{x-2}{x^2-5x+6}$				?			

Answer:

x	1.9	1.99	1.999	2	2.001	2.01	2.1
$\dfrac{x-2}{x^2-5x+6}$	.52632	.50251	.50025	?	.49975	.49751	.47619

Estimate of limit:

$$\lim_{x \to 2} \frac{x-2}{x^2 - 5x + 6} = 0.5$$

3. a) Give the definition of instantaneous rate of change of a function $f(x)$ at a point $x = c$.

$$f'(c) =$$

Answer:

$$f'(c) = \lim_{h \to 0} \frac{f(c+h) - f(c)}{h} \text{, if it exists.}$$

b) Find the instantaneous rate of change of $f(x) = x^2 + 2$ at $x = 1$ using the definition just asked for.

Answer:

$$f'(1) = \lim_{h \to 0} \frac{f(1+h) - f(1)}{h} = \lim_{h \to 0} \frac{[(1+h)^2 + 2] - [1^2 + 2]}{h} = \lim_{h \to 0} \frac{[1 + 2h + h^2 + 2] - 3}{h} =$$

$$\lim_{h \to 0} \frac{2h + h^2}{h} = \lim_{h \to 0} (2 + h) = 2 + 0 = 2$$

4. Find the equation of the line tangent to $y = 3x^2 + 2x + 1$ at the point $(2,17)$.

Differentiate: $y' = 6x + 2$. The slope of the tangent line at $(2,17)$ is y' evaluated at $x = 2$: $m = 6(2) + 2 = 14$. Use the point- slope form of a line to find the equation of the tangent line. Slope-intercept form: $y = 14x - 11$.

Answer:

$$y - 17 = 14(x - 2)$$

5. Find the derivative of each.

a) $f(x) = \dfrac{x^3}{x^2 + 1}$

Answer:

Differentiate, using the quotient rule.

$$f'(x) = \frac{(x^2 + 1)(3x^2) - x^3(2x)}{(x^2 + 1)^2}$$

Simplify:

$$f'(x) = \frac{x^4 + 3x^2}{(x^2 + 1)^2}$$

b) $f(x) = (x^2 + 4)^{20}$

Answer:

Differentiate, using the chain rule.

$$f'(x) = 20(x^2 + 4)^{19}(2x)$$

Simplify:

$$f'(x) = 40x(x^2 + 4)^{19}$$

6. Find $\dfrac{dy}{dx}$

Answer:

a) $y = e^{x^2} \ln x$

Differentiate, using the product and chain rules.

$$\frac{dy}{dx} = e^{x^2}\left(\frac{1}{x}\right) + \ln x\,[e^{x^2}(2x)]$$

Simplify:

$$\frac{dy}{dx} = e^{x^2}\left[\frac{1}{x} + 2x\ln x\right]$$

b) $y = \ln(x^2 + 1)$

Differentiate, using the chain rule. Then simplify.

$$\frac{dy}{dx} = \frac{1}{x^2 + 1}(2x)$$

$$\frac{dy}{dx} = \frac{2x}{x^2 + 1}$$

7. The number of ladybugs in an experiment was given by

$$y = f(t) = 5.285 + 1.603t^{2.539}$$

where t is the time measured in days.

a) Find $f'(t)$.

Differentiate, using the power rule.

Answer:
   $f'(t) = 1.603(2.539t^{1}.539)$

b) Find $f'(2)$ and explain what this means.

Evaluate: $f'(2) = 11.827$.

Answer:
   The number of ladybugs in the experiment was increasing by about 11.8 lady-bugs per day on the second day.

8. If $C(x) = 20 + 0.01x^2$ is a cost equation, where $C(x)$ is the cost in dollars of x items, use calculus to estimate the cost of the 31st item.

Differentiate to find marginal cost.

Answer:
   $C'(x) = 0.02x$
   The cost of the 31st item is $C'(30) = 0.60$, or $0.60.

9. Let $f(x) = x^3 - 3x^2 + 1$

a) Find all critical values.

Answer:

Differentiate:
$$f'(x) = 3x^2 - 6x$$

Factor and solve $f'(x) = 0$ to find critical values:
$0 = 3x(x-2)$. Critical values: $x = 0, 2$.

b) Find the largest open interval(s) where $f(x)$ is increasing.

Answer:

The graph of $f(x)$ is increasing on open intervals where $f'(x) > 0$. Use a sign chart to show that $f'(x) > 0$ on $(-\infty,0)(2,\infty)$. Therefore, $f(x)$ is increasing on $(-\infty,0)(2,\infty)$.

c) Find all relative maxima.

Answer:

Since 0 is a critical value and $f(x)$ is increasing on $(-\infty,0)$ and decreasing on $(0,2)$, $f(0) = 1$ is a relative maximum.

d) Find all relative minima.

Answer:

Since 2 is a critical value and $f(x)$ is decreasing on $(0,2)$ and increasing on $(2,\infty)$, $f(2) = -3$ is a relative minimum.

10. Let $f(x) = 1 + 15x + 3x^2 - 2x^3$

a) Find the largest open interval(s) where $f(x)$ is concave up.

Find the second derivative: $f'(x) = 15 + 6x - 6x^2$, $f''(x) = 6 - 12x$
Factor and solve $f''(x) = 0$ to find possible x-values where concavity changes: $0 = 6-12x$.
Solution: $x = 1/2$.
The graph of $f(x)$ is concave up on open intervals where $f''(x) > 0$. Use a sign chart to show that $f''(x) > 0$ on $(-\infty,\frac{1}{2})$.

Answer:

Therefore, $f(x)$ is concave up on $(-\infty,\frac{1}{2})$

b) Find all inflection points.

Answer:

The function $f(x)$ changes concavity at $x = 1/2$; therefore, $(1/2,9)$ is an inflection point.

11. Find the location of all absolute maxima and minima of $f(x) = x^3 - 3x^2 + 2$ on $[1,4]$.

Differentiate and find critical values: $f'(x) = 3x^3 - 6x$, $0 = 3x\,(x - 2)$, critical values are $x = 0, 2$. Evaluate the function $f(x)$ at the endpoints of the closed interval and any critical values within the interval. $f(1) = 0$, $f(2) = -2$, $f(4) = 18$.

Answer:

By the extreme value theorem, the absolute maximum must be 18 and the absolute minimum is −2.

12. Find the antiderivative.

$$\int \left(3e^x + \frac{4}{x} \right) dx$$

Integrate directly.

Answer:

$3e^x + 4\ln|x| + C$

13. Find the indefinite integral.

$$\int 10x(x^2 + 2)^5 dx$$

Integrate by substitution.

Answer:

Let $u = x^2 + 2$. Then $du = 2xdu$, so $dx = \dfrac{du}{2x}$

Using substitution, the integral becomes $\int 10x(u^5)\dfrac{du}{2x}$

which simplifies to $\int 5u^5\,du$

Integrate.

$$5\left(\frac{1}{6}u^6 \right) + C$$

Use $u = x^2 + 2$ to resubstitute and simplify.

$$\frac{5}{6}(x^2 + 2)^6 + C$$

14. Evaluate the definite integrals using the Fundamental Theorem of Calculus.

a) $\int_0^1 (6x^2 + 1)dx$

Answer:

Integrate:

$$\left[6\left(\frac{1}{3}x^3 \right) + x \right]$$

Simplify: $[2x^3 + x]$

Use the Fundamental Theorem of Calculus to evaluate.

$[2(1)^3 + 1] - [2(0)^3 + 0] = 3$

b) $\int_0^4 \sqrt{2x+1}\, dx$

Integrate by substitution.

Answer:

Let u = $2x + 1$, so $du = 2dx$ and $dx = (1/2)du.$

Change the bounds of the integral: $x = 0$ changed to $u = 2(0) + 1 = 1$, $x = 4$ changed to $u = 2(4) + 1 = 9$.

using substitution, the integral becomes $\frac{1}{2}\int_1^9 \sqrt{u}\, du$

Integrate and simplify:

$$\frac{1}{2}\left(\frac{2}{3}u^{3/2} \right) = \frac{1}{3}u^{3/2}$$

Use the Fundamental Theorem of Calculus to evaluate.

$$\frac{1}{3}(9^{3/2} - 1^{3/2}) = \frac{26}{3}$$

15. Let $f(t) = x^2 + 2x + 1$ on $[0,2]$, and let $n = 4$.

a) Make a sketch that illustrates the left-hand sum, showing clearly the 4 rectangles and t_0, t_1, t_2, t_3, t_4.

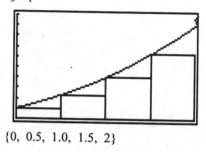

$\{0,\ 0.5,\ 1.0,\ 1.5,\ 2\}$

b) Use your calculator to find the left-hand and right-hand sums for $n = 10, 100,$ 500 (fill in the table).

n	LHS	RHS
10		
100		
500		

Answer:

n	LHS	RHS
10	7.88	9.48
100	8.5868	8.7468
500	8.650672	8.682672

Now use the preceding information to estimate the value of the definite integral.

$$\int_0^2 (x^2 + 2x + 1)\,dx \approx 8.7$$

16. A tank holding a polluting chemical breaks at the bottom and spills out at a rate given by $f'(t) = 400e^{-0.01t}$ gallons per day. How much spills out during the third day?

Answer:

Set up the integral.

$$\int_2^3 400e^{-0.01t}dt$$

Integrate by substitution.

Let $u = -0.01t$, so $du = -0.01dt$ and $dt = -100du$.

Change the bounds.

$t = 2$ changed to $u = -0.01(2) = -0.02$; $t = 3$ changed to $u = -0.03$.

Now substitute.

$$-40,000 \int_{-0.02}^{-0.03} e^u du = 40,000 \int_{-0.03}^{-0.02} e^u du$$

Finally, apply the Fundamental Theorem of Calculus.

$$40,000e^u \Big|_{-0.03}^{-0.02} = 40,000 (e^{-0.02} - e^{-0.03}) \approx 390$$

gallons spill out on the third day.

17. Given the two curves $y = x^2 - 2$ and $y = x$:

a) Sketch a graph of the region bounded by the two curves.

Answer:

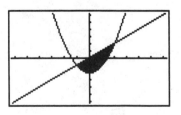

b) Write the definite integral(s) that give(s) the value of the area of the region.
DO NOT EVALUATE THE INTEGRAL(S).

Answer:

$$\int_{\text{left endpoint}}^{\text{right endpoint}} [\textit{top function} - \textit{bottom function}]dx = \int_{-1}^{2}[x - (x^2 - 2)]dx$$

PART FOUR

FOR YOUR REFERENCE

CHAPTER 23

COMMON MATHEMATICAL SYMBOLS

The following are the mathematical symbols used in this book.

+   "add" or "positive"

−   "subtract," "minus," or "negative"

×   "multiplied by" or "times"

*   "multiplied by" or "times"; also used with set notation to indicate exclusion of zero (see end of this list)

÷   "divided by" or "shared by"

/   "divided by" or "shared by"

±   "add or subtract"; "plus or minus"; "positive or negative"

=   "equals" or "is equal to"

≠   "does not equal" or "is not equal to"

≈   "is approximately equal to"

$\equiv$ "is identically equal to" or "is equivalent to" or "has the same value as"

$\not\equiv$ "is not equivalent to"

$<$ "is less than"

$\leq$ "is less than or equal to"

$>$ "is greater than"

$\geq$ "is greater than or equal to"

$\propto$ "varies as" or "is proportional to"

$\sim$ "varies as" or "is proportional to"

$[x]$ largest whole number that is not greater than x

$|x|$ absolute value of x with no sign attached

$\sqrt{x}$ "the square root of x"

$\sqrt[3]{x}$ "the cube root of x"

$\angle$ "angle"

$\parallel$ "is parallel to"

$\perp$ "is perpendicular to"

$\circ$ "degree"

$\{\}$ encloses a listed set

$()$, $[]$, $\{\}$ used in pairs to enclose an expression to be treated as a complete quantity or to be evaluated before the rest of the expression; for clarity, $()$ can be nested within $[]$, and $[]$ within $\{\}$

$\in$ "is a member of"

$\notin$ "is not a member of"

$\cap$ intersection of two sets

$\cup$ union of two sets

$\subset$ "is a subset of"

$\not\subset$ "is not a subset of"

$\supset$ "includes"

$\not\supset$ "does not include"

$\varnothing$ the null set

$\Rightarrow$ "implies"

$\Leftarrow$ "is implied by"

$\Leftrightarrow$ "implies and is implied by"

$\therefore$ "therefore"

∞ infinity

$f(x)$ "a function of x"

$\int f(x)dx$ indefinite integral

$\int_a^b f(x)dx$ definite integral

$\lim\limits_{x \to a} f(x) = b$ "the limit value of the function $f(x)$, as x approaches a, is b."

Δx "change in x"

Σ denotes summation

π pi (≈ 3.14159); also denotes a product

N the set of natural numbers and zero

N* the set of natural numbers

Z: the set of all integers

Z_+: the set of positive integers

Z_-: the set of negative integers

Z*: the set of all integers except zero

Q: the set of all rational numbers

Q_+: the set of positive rational numbers

Q_-: the set of negative rational numbers

Q*: the set of all rational numbers except zero

R: the set of all real numbers

R_+: the set of positive real numbers

R_-: the set of negative real numbers

R*: the set of all real numbers except zero

C: the set of all complex numbers

C*: the set of all complex numbers except zero

CHAPTER 24
SUGGESTED READING

Anderson, Robert F. V. *Introduction to Linear Algebra*. Holt, Reinhart and Winston, 1988.

Ayres, Frank, Jr., and Philip A. Schmidt, *Theory and Problems of College Mathematics*, 2d ed. McGraw-Hill, 1992.

Bobrow, Jerry. *Algebra I*. Cliffs Notes, 1994.

Bobrow, Jerry. *Algebra II*. Cliffs Notes, 1994.

Downing, Douglas. *Barron's Algebra the Easy Way*. Barron's, 1996.

Downing, Douglas. *Barron's Trigonometry the Easy Way*. Barron's, 1990.

Englefield, M. J. *Mathematical Methods for Engineering and Science Students*. Edward Arnold, 1987.

Farrand, Scott M., et al. *Calculus*. HBJ College and School Division, 1985.

Gullberg, Jan. *Mathematics: From the Birth of Numbers*. W. W. Norton, 1997.

Hahn, Alexander J. *Basic Calculus: From Archimedes to Newton to Its Role in Science*. Springer-Verlag, 1998.

Hamilton, A. G. *A First Course in Linear Algebra*. Cambridge University Press, 1987.

Hogben, Lancelot. *Mathematics for the Million*. 1937; reprint ed., W. W. Norton, 1993.

Kay, David A. *Trigonometry*. Cliffs Notes, 1994.

Larson, Roland E., with Robert P. Hostetler. *Calculus with Analytic Geometry*. 3d ed. Heath, 1986.

Lay, David C. *Linear Algebra and Its Applications*. Addison-Wesley, 1997.

Leff, Lawrence S. *College Algebra*. Barron's, 1995.

Lial, Margaret L., et al. *Trigonometry*. Addison-Wesley, 1997.

Mendelson, Elliot. *3000 Solved Problems in Calculus*. McGraw-Hill, 1992.

Miller, Bob. *Bob Miller's Algebra for the Clueless*. McGraw-Hill, 1999.

Munhe, Mustafa A., and David J. Foulis. *Algebra and Trigonometry with Applications*. Worth, 1982.

Paulos, John Allen. *Beyond Numeracy*. Knopf, 1991.

Pearson, Carl E., ed. *Handbook of Applied Mathematics*. 2d ed. Van Nostrand Reinhold, 1983.

Stewart, James. *Calculus*. Brooks/Cole, 1987.

Strang, Gilbert. *Linear Algebra and Its Applications*. HBJ College and School Division, 1988.

Tapson, Frank. *Barron's Mathematics Study Dictionary*. Barron's, 1996.

Thompson, J. E. *Algebra for the Practical Worker*. 4th ed. Van Nostrand Reinhold, 1982.

Thompson, J. E. *Calculus for the Practical Worker*. 4th ed. Van Nostrand Reinhold, 1983.

Thompson, Silvanus P., and Martin Gardner. *Calculus Made Easy*. St. Martin's Press, 1998.

Zandy, Bernard V. *Calculus*. Cliffs Notes, 1993.

Note: This index covers Part One: Preparing Yourself and Part Two: Study Guide, pages vii–75 of the text. Material in the sample exams, pages 80–278, is not indexed.